Advanced Technologies and Societal Change

This series covers monographs, both authored and edited, conference proceedings and novel engineering literature related to technology enabled solutions in the area of Humanitarian and Philanthropic empowerment. The series includes sustainable humanitarian research outcomes, engineering innovations, material related to sustainable and lasting impact on health related challenges, technology enabled solutions to fight disasters, improve quality of life and underserved community solutions broadly. Impactful solutions fit to be scaled, research socially fit to be adopted and focused communities with rehabilitation related technological outcomes get a place in this series. The series also publishes proceedings from reputed engineering and technology conferences related to solar, water, electricity, green energy, social technological implications and agricultural solutions apart from humanitarian technology and human centric community based solutions.

Major areas of submission/contribution into this series include, but not limited to: Humanitarian solutions enabled by green technologies, medical technology, photonics technology, artificial intelligence and machine learning approaches, IOT based solutions, smart manufacturing solutions, smart industrial electronics, smart hospitals, robotics enabled engineering solutions, spectroscopy based solutions and sensor technology, smart villages, smart agriculture, any other technology fulfilling Humanitarian cause and low cost solutions to improve quality of life.

Virender Kadyan · T. P. Singh · Chidiebere Ugwu
Editors

Deep Learning Technologies for the Sustainable Development Goals

Issues and Solutions in the Post-COVID Era

 Springer

Editors
Virender Kadyan
University of Petroleum and Energy Studies
Dehradun, Uttarakhand, India

T. P. Singh
University of Petroleum and Energy Studies
Dehradun, Uttarakhand, India

Chidiebere Ugwu
University of Port Harcourt
Port Harcourt, Rivers State, Nigeria

ISSN 2191-6853 ISSN 2191-6861 (electronic)
Advanced Technologies and Societal Change
ISBN 978-981-19-5725-3 ISBN 978-981-19-5723-9 (eBook)
https://doi.org/10.1007/978-981-19-5723-9

This Springer imprint is published by the registered company Springer Nature Singapore Pte Ltd.
The registered company address is: 152 Beach Road, #21-01/04 Gateway East, Singapore 189721,
Singapore

Preface

The emergence of artificial intelligence (AI) is shaping an ever-increasing range of sectors. For e.g., AI is believed to affect global productivity1, equality and inclusion2, environmental outcomes3, and several other areas, both in the short and long term4. Reported potential impacts of AI indicate both positive5 and negative6 impacts on Sustainable Development. Deep learning spanning from artificial intelligence (AI) is one of the technologies which has tremendous potential to revolutionize daily processes in various fields, leading humanity to an era of self-sufficiency and productivity. There is dire need to inspect AI applications that impacts UN Sustainable Development Goals, both positively and negatively. The goal is to promote the positive use of AI for Sustainable Development and to investigate on the negative impact of AI on Sustainable Development. It is time to discuss implications of how AI can either enable or inhibit the delivery of all 17 goals and 169 targets recognized in the 2030 Agenda for Sustainable Development. The focus of this book is to provide insights, how the deep learning techniques will impact the implementation of SDGs, what are their promises, limits, and the new challenges. This will also provide a publication avenue for researchers working on the impact deep learning approaches in implementing Sustainable Development Goals of UN. This book also covers the challenges, blockages, and opportunities in various applications of deep learning in SDGs. The main goal of this book is to present the comprehensive survey on the major applications and research-oriented articles based on deep learning techniques those are focused on Sustainable Development Goals. In particular, there is the need to extend research into deep learning and its broader application to many sectors and to assess its impact on achieving the Sustainable Development Goals (SDGs). The chapters in this book will help in finding the use of deep learning across all sections of SDGs. The rapid development of deep learning needs to be supported by the organizational insight and oversight necessary for AI-based technologies in general; here, we shall present and discuss implications of how deep learning enables the delivery agenda for Sustainable Development.

The book contains the collection of 16 chapters of research findings, reviews, and case studies by various experts in this volume. The book begins with an introduction of "How Deep Learning Can Help in Regulating the Subscription Economy to

Ensure Sustainable Consumption and Production Patterns (12th Goal of SDGs)." Each other in this book tried to bring his/her insight of SDG goals with different real-life applications.

Dehradun, India   Virender Kadyan
Dehradun, India   T. P. Singh
Port Harcourt, Nigeria   Chidiebere Ugwu

Acknowledgements

This SDG book would not be possible without support of my Springer's editorial team. The continues encouragement of them will help in completion of this book. Every book has a hard support of its editors and writers, so I would like to thank Chandrasekaran Arjunan and Loyola Dsilva from Springer team.

I remain grateful to the key supporter of this book my co-editors whose continuous help in timely completion of this assignment. Consequently, the success of this is not possible without the support of each and every reviewers who provided valuable input to each chapter writers. Dr. Amitoj Singh and Dr. Anupam Singh deservers special acknowledgment who has inserted his valuable thought in shaping of this book. Finally, I would like to thank each one of the chapter contributors without whom this book is not possible. So I am heartfully thank to each one of them for their hardwork and determinations.

Contents

Contributors

Abhishek Badholia Department of Computer Science and Engineering, MATS University, Raipur, Chhattisgarh, India

Gurpreet Singh Chhabra Department of Computer Science and Engineering, Gandhi Institute of Technology and Management, Visakhapatnam, Andhra Pradesh, India

Mridul Dharwal School of Business Studies, Sharda University, Greater Noida, India

Vishakha Goyal Sharda University, Greater Noida, India; Vivekananda College, Delhi University, New Delhi, India

Priya Gupta Atal Bihari Vajpayee School of Management and Entrepreneurship, Jawaharlal Nehru University, New Delhi, India

Shaurya Gupta School of Computer Science, University of Petroleum and Energy Studies, Dehradun, India

Anjali Jain Department of Electrical and Electronics Engineering, Amity University Uttar Pradesh, Noida, India

Avita Katal School of Computer Science, University of Petroleum and Energy Studies, Dehradun, Uttarakhand, India

Ashish Kumar School of Computer Science, Engineering and Technology, Bennett University, Uttar Pradesh, Greater Noida, India

Rintu Nath Vigyan Prasar, Department of Science and Technology, AI Building, New Delhi, India

Sylesh Nechully Haris Al Afaq LLC, Abudhabi, United Arab Emirates

Shailendra Kumar Pokhriyal School of Business, Himalayiya University, Doiwala, India

Ajay Prasad University of Petroleum and Energy Studies, Dehradun, India

Deepika Sachdev Application and Analytics, South East Asia Nokia Software, Singapore, Singapore

Anurag Sharma Department of Computer Science and Engineering, MATS University, Raipur, Chhattisgarh, India

Yogesh Sharma Atal Bihari Vajpayee School of Management and Entrepreneurship, Jawaharlal Nehru University, New Delhi, India

Vinod Kumar Shukla Department of Engineering and Architecture, Amity University, Dubai, UAE

Rajeev Sijariya Atal Bihari Vajpayee School of Management and Entrepreneurship, Jawaharlal Nehru University, New Delhi, India

Amit Singh School of Computer Science, University of Petroleum and Energy Studies, Dehradun, India

R. K. Singhal Dr A P J Abdul Kalam Technical University, Lucknow, India

Manoj Kumar Srivastava Management Development Institute, Gurgaon, India

Vishal Srivastava Dr A P J Abdul Kalam Technical University, Lucknow, India

Sai Shrinvas Sundaram Aptitude Global, Hyderabad, India

Ujjwal Department of Electrical and Electronics Engineering, Amity University Uttar Pradesh, Noida, India

Amit Verma School of Computer Science, UPES, Dehradun, Uttarakhand, India

Jyotsna Verma Department of Computer Science, The ICFAI University, Jaipur, India

Neelam Verma Department of Electrical and Electronics Engineering, Amity University Uttar Pradesh, Noida, India

Sourabh Singh Verma Manipal University, Jaipur, India

Vijayant Verma Department of Computer Science and Engineering, MATS University, Raipur, Chhattisgarh, India

Rubeena Vohra Bharati Vidyapeeth's College of Engineering, New Delhi, India

Sonali Vyas School of Computer Science, University of Petroleum and Energy Studies, Dehradun, India

Chapter 1
How Deep Learning Can Help in Regulating the Subscription Economy to Ensure Sustainable Consumption and Production Patterns (12th Goal of SDGs)

Yogesh Sharma, Rajeev Sijariya, and Priya Gupta

Introduction

If we look at the history of humankind and compare ourselves with them, we know we have come a long way. We are rapidly evolving and expanding [1], and due to the limited size of our planet, we are looking for signs of life on the other neighbouring planets. Billionaires like Jeff Bezos and Richard Branson have already made their first flight to space on Blue Origin and Virgin Galactic respectively[1] with an aspiration to make it possible for the public to experience space as well. We are moving towards the fourth industrial revolution, popularly known as Industry 4.0 [3], wherein companies are heavily investing in automation and data exchange (Fig. 1.1). We want to make our traditional factories into "smart factories", which will use technologies like the Internet of Things (IoT), artificial intelligence (AI), and cloud computing to manage the manufacturing operations faster and more efficiently.

Why Do We Need Faster and Efficient Manufacturing?

To cater to the consumption of a growing population worldwide, the countries are focusing on increasing their GDP growth. The global GDP, for instance, grew by

---

[1] Amos [2].

Y. Sharma (✉) · R. Sijariya · P. Gupta
Atal Bihari Vajpayee School of Management and Entrepreneurship, Jawaharlal Nehru University, New Delhi, India
e-mail: yogesh.ysharma93@gmail.com

© The Author(s), under exclusive license to Springer Nature Singapore Pte Ltd. 2023
V. Kadyan et al. (eds.), *Deep Learning Technologies for the Sustainable Development Goals*, Advanced Technologies and Societal Change,
https://doi.org/10.1007/978-981-19-5723-9_1

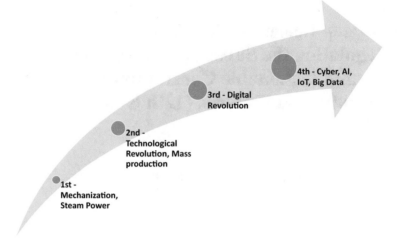

Fig. 1.1 Stages of Industrial Revolution. *Source* "The Fourth Industrial Revolution|Essay by Klaus Schwab|Britannica," 2021

173.8% in two decades from 1988 to 2018 (World Bank Data).[2] According to a Credit Suisse report,[3] there has also been an increase in global wealth, around $360 trillion in 2019. This has led to a chain reaction in our economy, and we are rapidly consuming natural resources by transferring them into goods and services to satisfy our needs and requirements [5].

During COVID-19, the world economy came to a halt due to the global lockdown [6]. Our consumption pattern also changed drastically [7, 8]. While the consumption of outdoor activities such as travelling, using transportation, sports, and leisure activities decreased, the consumption of indoor activities such as remote working, Internet usage, entertainment, online shopping, online streaming subscription, etc., increased significantly. Due to the vast potential, differentiation from competitors, and declining ownership trend, the subscription offering has emerged as the way forward to sell services and products in recent years. Moving ahead from the traditional pricing strategies of "penetration pricing" and "price skimming" [9], companies are now offering "Freemium" [10], wherein the service provider allows the consumer to experience the premium features for a certain period before purchasing the subscription services. To lure the customers and make them habitual of the subscription, the companies are offering potential customers with unlimited video content, music, Internet, calling, gym, food (in lounge), etc. Due to such unlimited offerings, our consumption is increasing rapidly, which is putting pressure on our natural resources.

---

[2] World Bank Group, "World Bank national accounts data, and OECD National Accounts data files, GDP (current US$)|Data", available at: https://data.worldbank.org/indicator/NY.GDP.MKTP.CD?end=2018&start=1989.

[3] Credit Suisse Group AG [4].

To ensure consumption in a regulated manner, the United Nations in its 12th goal of the Sustainable Development Goals (SDGs) talks about "Ensure sustainable consumption and production patterns".[4] Before COVID-19, the world has been using natural resources rapidly, in an unsustainable manner. To measure the impact, the United Nations has defined material footprint as "an attribution of global material extraction to domestic final demand of a country. The total material footprint is the sum of the material footprint for biomass, fossil fuels, metal ores, and non-metal ores". According to the data given by the UN, the Global Material Footprint (GMF) increased from 73.2 to 85.9 billion tonnes in just seven years (i.e. 2010–17), which is about a 17.34% increase. Electronic waste also grew by 38% and less than 20% of it has been recycled between 2010 and 19. Footprints or consumption-based indicators are aimed at behavioural change, and they can help in increasing awareness about the sustainable consumption [11].

There have been various business models introduced, such as product–service system (PSS), comprising of servitization and productization (discussed later in the chapter); recycling; green products and services and many more. But each of these models has its risks and returns. Another method that we know is reusing of the products, more specifically sharing. This idea is not new; we are aware of the barter system [12], and it was practised for centuries before the currency was invented. Back then, traders used to exchange goods in return for other goods. We do not do that anymore because now we have mass manufacturers and wealth (ATM, digital wallet, and online banking) to shop in the mall. Online shopping has captured those who could not take time to go out shopping by providing them, at their fingertips, with a shopping experience that they never experienced before [13]. The COVID-19 lockdown has acted as a catalyst in this situation. Even though many organizations and countries are talking about sustainable solutions for business growth, the impact is still less as compared to the immense pressure on our natural resources and leads to some bigger question which is raised and answered in the next section.

How Long Will We Be Able to Rely on Our Natural Resources for Our Rapid Consumption Demand?

Well, not for very long. Our consumption of resources is faster than resource replenishment which is leading to resource depletion. According to the Big Issue UK,[5] over £3.5 trillion worth of resources is idle worldwide. They reported empty houses, parked cars in driveways, food wastage and unused clothes, tools and types of machinery. And it is not only limited to tangible products but the services we consume

---

[4] United Nations—Department of Economic and Social Affairs—Sustainable Development. (2021), available at: https://sdgs.un.org/goals/goal12.

[5] The Big Issue [14].

at home with the help of the Internet. We are aware that the natural resources are rapidly exhausting [15] and will soon be on the verge of extinction. The long-term impact of under and overconsumption of subscription services is yet to be explored.

Subscription Services

Every other website and YouTube content creator asks you to subscribe to their mailers, notifications, or channel. Subscription services have grown rapidly, an area which was earlier dominated by SaaS (software as service) companies. As per the study by Gartner[6] in 2019, about 75% of the organizations will offer subscription services by 2023, out of which only 20% organizations will be able to grow their customer retention [17]. While the customers prefer subscription services due to convenience, less cost (compared to ownership), and customization, the organizations enjoy the certainty and predictable revenues.

Subscription services nowadays are based on diverse set of offerings. Core products are now being offered with bundle of services, as a value addition, termed as servitization [18] and core services are offered with bundle of products, termed as Productization [19]. Both ways, it is leading to the "product–service system" [20–24]. As compared to products, product–service system (PSS) can create values (tangible and intangible) by delivery customized solutions to the customers [25].

For understanding various types of PSS, Clayton et al. [26] categorized them as:

1. Pure products
2. Product supported with service
3. Product and service
4. Service supported with product
5. Pure service.

The PSS model has caught eyes of academicians and businesses as an effective strategy in reducing consumption of natural resources and improving the overall environmental sustainability [27]. Subscription-based businesses can take the leverage of the PSS model. Instead of just selling their product or service and cutting the umbilical cord of further interaction with their customers, companies have understood how important it is to keep the connection alive by offering subscriptions. Subscription services can also be segmented as follows.

Mode of Delivery (Online/Offline/Mix)

Subscription services can be offered online, offline, and a mix of both [28]. Online mode will include online learning, Netflix, Amazon Prime, and other online streaming

---

[6] Gartner, Inc. [16].

applications, offline subscription services can be newspaper delivery, milk delivery, car-washing and more, and the mix will include services like online–offline banking, monthly subscription to Uber and Ola in India, wherein you are booking, making payment, and giving feedback online, but using the taxi service offline.

Payment System (Pay as Use/Periodic)

Depending on the usage, the payment system is also flexible [29]. A customer can either pay as they use or choose a monthly, yearly, or even lifetime subscription. The example for pay as a customer use can be electricity or water subscription, a monthly subscription can be a gym membership, a yearly subscription can be of telecommunication services, and a lifetime subscription of some institution.

Limited and Unlimited Subscription

Subscription services can also be differentiated on the basis of how they are being offered, i.e. limited or unlimited [30]. Examples of limited services can be a gym membership that does not allow its customers to visit the gym more than once on the same day or a newspaper delivery subscription, giving its customers only one paper per day. In case of unlimited subscription offering, the examples are vast, from the unlimited buffet at airport lounges to unlimited Internet and domestic calling. Online streaming services also allow their customers to watch as much content as and whenever they want. Unlimited offering attracts more customers if they compare the same services with limited offerings. In India, when Reliance Jio announced free unlimited calling and Internet to its users in 2016 [31], a huge number of Indians started queuing up outside Jio stores to subscribe to their services. They acquired about 16 million subscribers in the first month of their launch. It was a game-changer offering, leading other players such as Bharti Airtel, Idea, and Vodafone to offer similar unlimited packages. It took some time for the competitors to come back with an offer, meanwhile, Jio enjoyed the first movers' advantage. Both consumers and companies are focusing on optimizing their self-interests. While the companies prefer unlimited offerings to enjoy a large customer base and profits, the customers enjoy the unlimited offers and consume more than they require. This is due to the comparison of value for the money paid, wherein the customer expects to get the maximum value out of the price they are paying for that subscription. Our screen time on mobiles, TVs, and laptops has increased significantly. The concept of binge-watching and online shopping has been widely researched. Consumers' habits are changing, and they are becoming digitally overloaded consumers. The side-effects of such overloaded consumption on customers, companies, environment, and digital infrastructure are discussed below.

Customers

The companies offer unlimited subscription to capture a larger market share. They analyse how much a customer can consume on a particular day. For example, how long can you watch movies/shows on Netflix in a day? Maybe 5–10 h. In a buffet, how much can a customer eat? Maybe 2–3 plates of food. In a gym, for how many hours can a customer work out maximum? Maybe 3–4 h. A customer may reach the total utility [32] of that subscription service and not use it for some days. In the long term, the customer may also feel that he is now bored exploring the service or does not require unlimited service any further, leading to unsubscribing and customer churn [33].

Companies

Sometimes, it is not easy for the companies providing unlimited services to produce in such large quantities or keep refreshing their contents. Netflix has to come up with new content, airport lounges have to be ready to refill food and other companies have to ensure all efforts so that their customers can enjoy seamless unlimited services without any interruptions. They work under pressure, as they know their customer might switch to the competitors, who are offering similar subscription packages. To stop customers from doing this, some companies even set higher initial fees of subscription so that customers stick for some fixed duration [34].

Environment

Customers are consuming more than what they require, and companies are producing continuously well in advance so that they keep their promise of providing unlimited offerings and the customer does not suffer due to waiting or unavailability. Consumption has increased dramatically, and so is the production to meet the consumption demands [7, 8]. In this cycle, we have put a lot of pressure on natural resources. There is no systematic regulation to monitor the consumption of the customers and the production of the companies.

Digital Infrastructure

In 2014, a landmark judgement was passed by the European Court of Justice (ECJ), concerning with the "right to be forgotten", which allows EU citizens to ask the search engines to remove the "inadequate, irrelevant, or excessive" content related to them [35]. On International E-waste Day 2020, the WEEE forum came up with a thoughtful report discussing how Internet waste is generated and its impact on the environment. Various precious metals and ores are used to create different electrical and electronic devices. With the emergence of IoT, the consumption of these metals will increase further, leading to an increase in e-waste [36]. Internet consumption will also increase rapidly. To store and transfer such a mammoth amount of data, infrastructure including antennas, servers, data warehouses, etc. is required. Even if we are texting, reading, or playing music, some amount of energy is consumed, and some data is used or stored somewhere. In the race of developing IoT solutions, our non-electrical objects such as furniture will also become smart with the help of Internet. Still, at the same time, these objects will also consume some energy. Such non-electrical objects powered by IoT will start contributing adversely to the carbon footprint. IoT devices will also be powered by some Internet device or cellular network, which will also consume some energy due to the transfer of data. To reduce the energy consumption due to the infrastructural requirements of the 5G technology in the telecom sector, Sun et al. [37] developed a neural network to reduce the energy consumed and ensuring customer satisfaction in terms of the quality of the cellular network. Similarly, Sekaran et al. [38] also proposed a model ensuring cost and energy efficiency in 6G technology. A lot of studies have been conducted on e-waste [39–46] and on managing e-waste using IoT [47–50]. But how to manage Internet waste is still being explored.

Role of Artificial Intelligence in Subscription Business

The concept of artificial intelligence (AI), popularly known as AI, roots from the 1950s but has gained popularity recently due to the advancement in the hardware and software domain, allowing data scientists to work on algorithms. It was developed to make decisions like humans by taking a heuristic approach to complex problems. For example, AI can automatically classify e-books based on the table of contents and show the relevant results [51]. But to solve such problems, it requires data just like humans need experience to make better decisions. Any company that has data is the king. Data makes profits.

While using social media and other online shopping websites, most of us agree to the terms and conditions without reading them, as we are unable to comprehend the legal terminologies. By taking this leverage, many data scientists are analysing a large set of diverse past and real-time data to develop complex algorithms based on trends, identifying customer preferences, and customizing offerings accordingly to target their customers effectively. Machine learning techniques can be employed in online advertising by analysing online customer responses [52]. Today's customer is a digital customer, as he/she is surrounded by various digital touchpoints, such as social media, mobile phone apps, fitness gears, digital advertisements, chatbots, voice recognition (Siri and Alexa), text recognition, IoT, virtual reality, etc. These are the various touchpoints which are driven by AI. The concept of sentiment analysis has opened doors for market researchers to get deeper into consumer insights. The applicability is not only limited to marketing, but researchers have also used sentiment analysis to predict election results. Sankar et al. [53] suggest that to ensure efficient sentiment analysis, machine learning can be used to filter out primary texts that can be fed into a convolutional neural network (CNN) for further processing.

AI is also helpful in detecting banking frauds, wherein it captures usage patterns of their customers and identifies suspicious transactions and alert the bank. By using machine learning algorithms, the banks can predict customers' intention to cancel their credit card subscription [54].

Due to the growing population and climate change, there is immense pressure on the agricultural sector to ensure an optimum food supply. The concept of artificial intelligence can be applied in agriculture. By predicting the optimal frequency of spraying fertilizers based on the crop, the farmers can increase their yield and save costs, hence promoting UN SDG of responsible consumption and production and ensuring zero hunger [55]. Similarly, irrigation management can also be more effective with the help of convolutional neural networks (CNN). It can predict water requirements so that water resources can be managed in advance [56].

The algorithms used in AI also consume some energy. Researchers advocate developing algorithms that consume less processing time and energy consumption. Henderson et al. [57] proposed a framework to track and report real-time carbon emission and energy consumption in machine learning experiments and calling for energy-efficient algorithms.

Evolution of Deep Learning to Support AI

The deep learning concept was introduced to overcome the limitations of machine learning. Due to the availability of large datasets and advances in AI, deep learning methods can solve not only routine problems but also complex problems exceeding the human level. The artificial neural network (ANN) used in deep learning algorithms consists of multiple layers [58], making it process complex problems and predict the results with accuracy. The multiple layers can be categorized into three parts, the first being the input layer, followed by the second part of "N" hidden layers and the third part being the output layer.

The concept of deep learning became popular due to its more extensive applicability in the area of recognizing images, speech and text or handwriting. For example, by using the deep learning model, Omidvar et al. [59] were able to predict the quality of the online news. To overview and identify the on-going research areas incorporating deep learning and subscription, Fig. 1.2 shows the concepts and associated terminology related to deep learning based on the extensive literature available.

Fig. 1.2 Word cloud of deep learning and associated terms. *Source* Authors' compilation of the literature on deep learning and subscription available on Scopus

Paradox: Deep Learning and Sustainable Consumption

Artificial intelligence (AI) and IoT accompaniment each other. To implement AI, the concept of machine learning (ML) evolved and to overcome the limitations of ML [60], the idea of deep learning came into existence. Some of the applications of deep learning can be seen in self-driving cars [61], fraud detection [62], virtual assistant and chatbots [63], visual recognition [64], voice recognition [65], colourization of black and white images and videos [66], language translation [67], election predictions [68], etc. On the one hand, we can agree that innovations using deep learning technologies are helping us in achieving good quality of life and making our lives easier, but on the other hand, they are also consuming big data. A huge amount of data is transmitted using the Internet, processed, and compiled using electrical signals and stored in some electronic devices using components made up of precious metals.

> Are we saying that the more technologically advanced we are becoming, the more harm we provide to the environment?

One may argue that we may use deep learning and IoT to regulate and efficiently manage our e-waste and Internet waste. This leads to another question,

> Can we develop our deep learning algorithms smart enough to make decisions considering the impact on the environment and reducing carbon and material footprints?

There is a pool of research available exploring the area of Green Supply Chain Management (GSCM) focusing on making manufacturing sustainable and Green Internet of Things (G-IoT) focusing on waste management using IoT. The role of deep learning is very crucial in managing subscription services.

Deep Learning as a Solution to Regulate Subscription Economy and Ensuring Sustainable Consumption

The role of AI and big data in developing sustainable consumption habits and the green behaviour of the consumer is still in the exploratory stage and requires further research [69]. To facilitate the UN SDG ensuring sustainable consumption, the products and services should be designed in a way that they have sustainability attributes, and the customer choices should be regulated with limited products and services which promotes sustainability and reduces pressure on natural resources [70].

The role of the PSS has been widely discussed due to its environmental performance and promoting reduced consumption of natural resources. When any business offers a combination of products and services, the customer is not the real owner of the product-service but uses the offering for a period and pays only for the consumption. The manufacturer takes care of the repair and maintenance in case of any fault. By employing deep learning algorithms, the manufacturers can monitor the product's health, detect faults by getting alerts at an early stage and react promptly, thereby saving on costs to repair and ensuring the long life of the product [27]. Following are the application of deep learning in different subscription services to regulate consumption.

Reducing Customer Churn in Subscription of OTT Platforms

In the subscription model, the aim is to maintain a long-term relationship with the existing customers. Therefore, any subscription-based business must minimize their customer churn [33], i.e. reduce the number of unsubscribing customers. Churn could be due to various reasons, depending on the type of business. For example, considering usage frequency and payment frequency, we can predict which customer is likely to stay longer. Under and overconsumption are the two extremes that any company should avoid. Suppose a customer has subscribed to OTT (over-the-top) services for a year, but they hardly get time to watch it. The other case could be binge-watching, wherein the customer explores almost every series on the OTT platform and does not have anything new to explore. Both issues may lead to customer churn. Using deep learning tools to predict customer churn, in this case, is not difficult as the companies have access to real-time data of each customer and their interaction with the OTT platform.

Similarly, with the help of vector embedding in deep learning, telecom companies can predict whether a customer will continue their subscription membership or not [71]. Deep learning can also help churn prediction in online casual games' subscriptions [72]. The implementation of deep learning will not only help the companies to reduce churn and avoid under or overconsumption but also lessen the burden on our natural resources, creating a path for sustainable consumption.

Collaborating Internet of Vehicle (IoV) and Deep Learning in Vehicle Subscription

The vehicle subscription market is growing rapidly, due to the flexibility and freedom from ownership and other liabilities. COVID-19 also acted as a catalyst for increasing the demand for private vehicles, as people would like to avoid using public transport [73]. Many companies in Europe, the USA, Canada, India, China, etc., offer these services. Such companies have tied up with manufacturers, for example, in India, Revv (India) is working with Hyundai, Zoomcar with Volkswagen, Nissan, Mahindra & Mahindra, and Renault and Myles is working with MG Motors, Toyota, and Maruti Suzuki. The subscription prices vary depending on the model and duration. In the case of Maruti Suzuki, the price ranges between $156 and $428 and duration ranging between 12 and 48 months. Luxury carmakers are also stepping into the subscription-based business to target those customers who aspire to these brands but do not want to invest huge sums of money. In 2020, Volvo also announced offering subscriptions on their cars in India.[7] This model promotes sustainable consumption as customers will subscribe based on requirements and return the car, so instead of the car being parked in a garage, it will be used by the next customer. While basic features like a speed limiter and eco mode driving are already available to ensure optimal fuel consumption, the implementation of deep learning can further enhance the performance of such models. Technologies such as the Internet of Vehicles (IoV) [75] can also be deployed to support green mobility by ensuring reduced carbon emissions. The IoV can be combined with 5G technology and deep reinforcement learning to minimize energy consumption and promote green IoV [76].

Minimizing Food Wastage and Global Hunger

Food waste and global hunger are the two extremes that exist simultaneously. Worldwide hunger is a severe issue, and the COVID-19 pandemic has worsened the situation even further [77]. Suppose an individual has subscribed to the frequent flyer program of any airline or alliance. In that case, they get access to a variety of unlimited food and beverages at the airport lounges. In an unlimited buffet, a person can take small portions of food depending upon the hunger level. But most of us take more than what we usually consume and leave the remaining food on the plates. Therefore, a framework to track the material flow and alert the food waste management systems can help reducing food waste [78]. The deep learning algorithms can be developed to analyse the contents in dustbins, CCTV footage of passengers in the lounge and their eating habits. By doing so, it can suggest optimum portion size, quantity of utensils and send alerts the concerned food waste management systems to distribute the remaining food to the NGOs. A case study on "Reducing food waste through

---

[7] HT Auto DeskHT Auto Desk [74].

computer vision technology" by Winnow[8] is about the food wastage which is a major problem in the hospitality industry. Much food gets wasted during transit, storage, perishability, and human consumption. Machine learning can help manage the food supply chain by developing a proper system to forecast future demand for food, inventory management, dynamic pricing based on the expiry date of the product and predicting weather and transportation disruptions. By using the concept of AI and machine learning, winnow provides food waste solutions to various commercial kitchens around the world. They have developed an AI-based tool, winnow vision, that helps chefs track their food wastage. The tool trains itself by processing the images of food thrown in dustbins over a period of time, presenting that data to the kitchen team to measure and manage food waste during operations in the future.

Convolutional Neural Network (CNN) to Help Shoppers in Identifying Green Products

Customers can be responsible shoppers even when they are not aware of the issues such as carbon emissions, footprints, and other environmental impacts. Using the concept of image processing with the help convolutional neural network (CNN), customers can fetch information about carbon emission by various products as they shop, promoting green consumption behaviour [15]. Asikis et al. [80] discuss how a personal shopping assistant in the form of a smartphone app can be used to recommend its user about certain sustainable products based on customer reviews and expert opinions available online.

Regulating the Energy and Water Consumption

The recent studies have employed deep learning in two areas, i.e. forecasting energy demand and reducing energy consumption. The study by Lenherr et al. [81] points out that deep learning models focuses on the accuracy of the predicted results, ignoring environmental costs. As a solution, they suggested metrics that include energy consumption measurements.

The consumption of electricity can be predicted using deep learning and dynamic pricing can be employed according to the demand to regulate consumption. By using IoT sensors, the data on energy consumption can be gathered to predict future energy demand [82]. In the electricity subscription, the provider can forecast the load with the help of a smart grid connected to IoT smart devices at individual homes to gather demand data and change the price accordingly [83]. For example, during peak hours,

---

[8] Food Waste Technology|Winnow [79].

when the demand is high, the electricity provider may use surge pricing, i.e. to increase the price per unit consumption, making the households consume electricity responsibly. Kant and Sangwan [84], with the help of AI, developed a predictive model to find out the power consumption in the machining process.

Deep learning can also be used to estimate water demand and consumption, predict infrastructure maintenance, assess water quality, monitor reservoirs, and predict floods and draughts, all of which are important aspects of sustainable water consumption [85]. Bejarano et al. [86] proposes a Smart Water Prediction System (SWaP) that helps in predicting future water consumption demand. The pricing can be dynamic, based on the availability and demand of water, to regulate the consumption of customers. There is a case on "machine learning helps to cut energy consumption by 40% in Google's data centres"[9] about Google LLC which is a popular and leading multinational company providing a diverse range of Internet-based services like online advertising, search engine and cloud computing, marking its presence and popularity across users around the globe. On average, Google's search engine processes 63,000[10] queries per second to show the results, requiring a vast amount of data and data centres to process and store information. Data centres can consume much electricity if they are not properly optimized and managed. To reduce energy consumption in the data centres, Google collaborated with DeepMind[11] to find a solution. By using AI and machine learning algorithms to predict future energy consumption, they developed an efficient way to cool down the data centres, resulting in a 40% decrease in energy consumption, hence contributing towards reducing carbon footprint.

Another case study on "AI for Sustainable Water in Bengaluru".[12] Bengaluru city, popularly known as the "Silicon Valley of India", is located in the south part of India. Being an IT hub, it has been rapidly growing in terms of businesses and migrant workers. Like any other big city, it faces some challenges, and the most alarming one is the water crisis. According to World Resources Institute (WRI), Bengaluru is facing severe water shortages and heading towards "zero day", just like Cape Town. The Bengaluru Water Supply and Sewerage Board (BWSSB) manages the water supply in the city. They faced challenges in their operations due to contributing factors like physical leaks in the distribution system, inefficient distribution system, and commercial losses due to human error in recording or faulty metres. These challenges are hampering the long-term sustainability of the city and the environment. SmartTerra, a start-up offering water management solutions, developed an AI-based intelligence platform, allowing predictive and efficient operations. It analyses the water distribution losses, health of the network and efficiency of the distribution, thereby helping the urban cities to manage their water distribution operations smoothly.

---

[9] Sanu [87].

[10] Prater [88].

[11] DeepMind AI Reduces Google Data Centre Cooling Bill by 40%. (2016). Deepmind. https://deepmind.com/blog/article/deepmind-ai-reduces-google-data-centre-cooling-bill-40.

[12] AI for Sustainable Water in Bengaluru [89].

Conclusion

Discussions on Findings

The role of deep learning in the subscription economy is immense and offering subscription services in a way to ensure sustainable consumption is still unexplored. However, in this chapter, we discussed how we could ensure sustainable consumption under various subscription services with the help of deep learning by gathering examples from the literature. Yet, there is much more to explore. With the technological advancements, we must ensure how to implement those innovative ways in a sustainable manner so that we do not exploit natural resources.

This chapter reveals how rapidly increasing consumption adversely affects our natural resources, making it difficult for the resources to replenish. Achieving the sustainable development goals (SDGs) is essential to make our planet liveable for our future generations. The impact of under and overconsumption under subscription services should be studied at greater length, as it will allow us to determine and regulate the optimum level of consumption. While the companies are exploring and coming up with innovative subscription models, they should also assess the impact of their offering on the environment and resource consumption so that planet is not compromised with earning more profits. The scope of deep learning is vast and versatile due to its applicability in various fields, such as banking, finance, medical research, construction, transportation, and many other areas, but technological advancements should go hand in hand with ensuring a safe and healthy environment. Not only businesses but the customers should be equally held responsible. Reminders from the service provider about under or overconsumption to their customers can alert them about their consumption patterns. The goal is to make the customers as "responsible consumers" and companies as "responsible producers", only then we will be able to achieve the UN 12th SDG of sustainable consumption and production patterns.

Scope for Future Research

This chapter contributes by highlighting the various practical applications for implementing deep learning techniques in the area of subscription-based services and ensuring sustainable consumption. While reviewing the existing literature and secondary data, it was observed that while the studies on the role of deep learning in subscription-based services and on sustainable consumption are available in abundance, but there were no attempts to link the two concepts, i.e. sustainable consumption and subscription services, while exploring the role of deep learning. Researchers may use this chapter as a stepping stone to further investigate the advanced applications of deep learning in various sectors under the subscription economy while ensuring sustainable consumption and predicting the long-term impacts.

Implications for Practitioners

Due to the rapid and unregulated surge in consumption of resources, it has become essential for technology-driven growth to ensure optimum and responsible usage of resources. Offering unlimited offers in subscription-based businesses may help acquire a large customer base in the short run, but it may not be a viable solution in the long run. The focus of subscription businesses is to reduce customer churn and maintain a long-term relationship with the existing customers, as acquiring new customers is expensive. Deep learning is a helpful way to classify customers based on their activity and predict customer churn. Further, it can help in detecting unusual activities, fraud users, and payments. Deep learning can also learn how much consumers are consuming the subscription service, and based on that, it can issue reminders on overuse, like iPhone issues weekly screen time analysis. The essence is to ensure balanced consumption, which will contribute to sustainable consumption.

References

1. Nutkiewicz, A., Jain, R.K.: Exploring the integration of simulation and deep learning models for urban building energy modeling and retrofit analysis. In: Proceedings of Building Simulation 2019: 16th Conference of IBPSA (2019). https://doi.org/10.26868/25222708.2019.210264
2. Amos, J.: Sir Richard Branson takes off on "extraordinary" space flight. BBC News (2021). https://www.bbc.com/news/science-environment-57790040
3. Lasi, H., Fettke, P., Kemper, H.G., Feld, T., Hoffmann, M.: Industry 4.0. Bus. Inf. Syst. Eng. **6**(4), 239–242 (2014). https://doi.org/10.1007/s12599-014-0334-4
4. Credit Suisse Group AG: Global Wealth Report 2019 (2019). Available at: https://www.credit-suisse.com/media/assets/corporate/docs/about-us/research/publications/global-wealth-report-2019-en.pdf
5. Zhuang, H., Zhang, J., Sivaparthipan, C.B., Muthu, B.A.: Sustainable smart city building construction methods. Sustainability **12**(12), 4947 (2020).https://doi.org/10.3390/su12124947
6. Muhammad, S., Long, X., Salman, M.: COVID-19 pandemic and environmental pollution: a blessing in disguise? Sci. Total Environ. **728**, 138820 (2020). https://doi.org/10.1016/j.scitotenv.2020.138820
7. Hasan, M.M., Roy, P., Sarkar, S., Khan, M.M.: Stock market prediction web service using deep learning by LSTM. In: 2021 IEEE 11th Annual Computing and Communication Workshop and Conference, CCWC 2021, pp. 180–183. Institute of Electrical and Electronics Engineers Inc. (2021) https://doi.org/10.1109/CCWC51732.2021.9375835
8. Hasan, S., Islam, M.A., Bodrud-Doza, M.: Crisis perception and consumption pattern during COVID-19: do demographic factors make differences? Heliyon **7**(5), e07141 (2021). https://doi.org/10.1016/j.heliyon.2021.e07141
9. Feng, S., Hu, X., Yang, A., Liu, J.: Pricing strategy for new products with presales. Math. Probl. Eng. **2019**, 1–13 (2019). https://doi.org/10.1155/2019/1287968
10. Oestreicher-Singer, G., Zalmanson, L.: Content or community? A digital business strategy for content providers in the social age. MIS Quart.: Manag. Inf. Syst. **37**(2), 591–616 (2013). https://doi.org/10.25300/MISQ/2013/37.2.12
11. Dawkins, E., Kløcker Larsen, R., André, K., Axelsson, K.: Do footprint indicators support learning about sustainable consumption among Swedish public officials? Ecol. Ind. **120**, 106846 (2021). https://doi.org/10.1016/j.ecolind.2020.106846

12. Liesch, P., Hill, R.S., Birch, D.: 13.6 pricing and countertrade (Pricing & Barter): contemporary issues in pricing. In: Proceedings of the 1995 World Marketing Congress, pp. 197–204 (2015). https://doi.org/10.1007/978-3-319-17311-5_32
13. Pham, Q., Tran, X., Misra, S., Maskeliūnas, R., Damaševičius, R.: Relationship between convenience, perceived value, and repurchase intention in online shopping in Vietnam. Sustainability **10**(2), 156 (2018). https://doi.org/10.3390/su10010156
14. The Big Issue: How the sharing economy can help us ride out the recession (2020). Available at: https://www.bigissue.com/opinion/how-the-sharing-economy-can-help-us-ride-out-the-recession/
15. Lu, W.Y., Chiu, M.C.: Apply deep learning image recognition technique to promote green consumer behavior. In: Advances in Transdisciplinary Engineering, vol. 10, pp. 205–213. IOS Press BV (2019). https://doi.org/10.3233/ATDE190125
16. Gartner, Inc.: Top 10 trends in digital commerce—smarter with Gartner (2019). Available at: https://www.gartner.com/smarterwithgartner/top-10-trends-in-digital-commerce/
17. Moore, S.: Top 10 trends in digital commerce—smarter with Gartner. Copyright (C) 2021 Gartner, Inc. All Rights Reserved (2019). https://www.gartner.com/smarterwithgartner/top-10-trends-in-digital-commerce/
18. Vandermerwe, S., Rada, J.: Servitization of business: adding value by adding services. Eur. Manag. J. **6**(4), 314–324 (1988). https://doi.org/10.1016/0263-2373(88)90033-3
19. Leoni, L.: Productisation as the reverse side of the servitisation strategy. Int. J. Bus. Environ. **10**(3), 247 (2019). https://doi.org/10.1504/ijbe.2019.097981
20. Baines, T.S., Lightfoot, H.W., Evans, S., Neely, A., Greenough, R., Peppard, J., Wilson, H.: State-of-the-art in product-service systems. Proc. Inst. Mech. Eng., Part B: J. Eng. Manuf. (2007). https://doi.org/10.1243/09544054JEM858
21. Goedkoop, M.J.: Product service systems, ecological and economic basics product service systems, ecological and economic basics. In: Report for Dutch Ministries of Environment and Economic Affairs (1999)
22. Manzini, E., Vezzoli, C., Clark, G.: Product-service systems: using an existing concept as a new approach to sustainability. J. Des. Res. (2001). https://doi.org/10.1504/jdr.2001.009811
23. Mont, O.K.: Clarifying the concept of product-service system. J. Clean. Prod. (2002). https://doi.org/10.1016/S0959-6526(01)00039-7
24. Tukker, A.: Eight types of product-service system: eight ways to sustainability? Experiences from suspronet. Bus. Strategy Environ. (2004). https://doi.org/10.1002/bse.414
25. Tukker, A., Tischner, U.: Product-services as a research field: past, present and future. Reflections from a decade of research. J. Cleaner Prod. (2006). https://doi.org/10.1016/j.jclepro.2006.01.022
26. Clayton, R.J., Backhouse, C.J., Dani, S.: Evaluating existing approaches to product-service system design: a comparison with industrial practice. J. Manuf. Technol. Manag. **23**(3), 272–298 (2012). https://doi.org/10.1108/17410381211217371
27. Ren, S., Zhang, Y., Sakao, T., Liu, Y., Cai, R.: An advanced operation mode with product-service system using lifecycle big data and deep learning. Int. J. Precis. Eng. Manuf. Green Technol. (2021). https://doi.org/10.1007/s40684-021-00354-3
28. Nguyen, G.D., Dejean, S., Moreau, F.: On the complementarity between online and offline music consumption: the case of free streaming. J. Cult. Econ. **38**(4), 315–330 (2013). https://doi.org/10.1007/s10824-013-9208-8
29. Xu, J., Li, H., Tayur, S.R.: Online-to-offline platform models. SSRN Electron. J. (2019). https://doi.org/10.2139/ssrn.3449744
30. Lu Wang, C., Zhang, Y., Richard Ye, L., Nguyen, D.-D.: Subscription to fee-based online services: what makes consumer pay for online content? J. Electron. Commer. Res. **6**, 304–311 (2005). Retrieved from http://www.isy.vcu.edu/~jsutherl/Info658/FFSINFO.pdf
31. Curwen, P.: Reliance Jio forces the Indian mobile market to restructure: a regular column on the information industries|emerald insight. Digital Policy Regul. Gov. (2017). https://doi.org/10.1108/DPRG

32. Alvino, L., Constantinides, E., Franco, M.: Towards a better understanding of consumer behavior: marginal utility as a parameter in neuromarketing research. Int. J. Mark. Stud. **10**(1), 90 (2018). https://doi.org/10.5539/ijms.v10n1p90
33. Deligiannis, A., Argyriou, C.: Int. J. Inf. Eng. Electron. Bus. (IJIEEB) **12**(4), 1 (2020). http://www.mecs-press.org/ijieeb/ijieeb-v12-n4/v12n4-1.html
34. Lee, S.H.: An exploration of initial purchase price dispersion and service-subscription duration. Sustainability **11**(9), 2481 (2019). https://doi.org/10.3390/su11092481
35. Bowcott, O.: "Right to be forgotten" by Google should apply only in EU, says court opinion. The Guardian (2019). https://www.theguardian.com/technology/2019/jan/10/right-to-be-for gotten-by-google-should-apply-only-in-eu-says-court
36. Awasthi, A.K., Cucchiella, F., D'Adamo, I., Li, J., Rosa, P., Terzi, S., Wei, G., Zeng, X.: Modelling the correlations of e-waste quantity with economic increase. Sci. Total Environ. **613–614**, 46–53 (2018). https://doi.org/10.1016/j.scitotenv.2017.08.288
37. Sun, G., Ayepah-Mensah, D., Xu, R., Boateng, G.O., Liu, G.: End-to-end CNN-based dueling deep Q-Network for autonomous cell activation in Cloud-RANs. J. Netw. Comput. Appl. **169** (2020). https://doi.org/10.1016/j.jnca.2020.102757
38. Sekaran, R., Ramachandran, M., Patan, R., Al-Turjman, F.: Multivariate regressive deep stochastic artificial learning for energy and cost efficient 6G communication. Sustain. Comput.: Inf. Syst. **30** (2021). https://doi.org/10.1016/j.suscom.2021.100522
39. Grant, K., Goldizen, F.C., Sly, P.D., Brune, M.N., Neira, M., van den Berg, M., Norman, R.E.: Health consequences of exposure to e-waste: a systematic review. Lancet Glob Health **1**(6) (2013). https://doi.org/10.1016/S2214-109X(13)70101-3
40. Kahhat, R., Kim, J., Xu, M., Allenby, B., Williams, E., Zhang, P.: Exploring e-waste management systems in the United States. Resour. Conserv. Recycl. (2008). https://doi.org/10.1016/j.resconrec.2008.03.002
41. Osibanjo, O., Nnorom, I.C.: The challenge of electronic waste (e-waste) management in developing countries. Waste Manage. Res. **25**(6), 489–501 (2007). https://doi.org/10.1177/073424 2X07082028
42. Perkins, D.N., Brune Drisse, M.N., Nxele, T., Sly, P.D.: E-waste: a global hazard. Ann. Glob. Health (2014). https://doi.org/10.1016/j.aogh.2014.10.001
43. Robinson, B.H.: E-waste: an assessment of global production and environmental impacts. Sci. Total Environ. (2009). https://doi.org/10.1016/j.scitotenv.2009.09.044
44. Terazono, A., Murakami, S., Abe, N., Inanc, B., Moriguchi, Y., Sakai, S.I., et al.: Current status and research on E-waste issues in Asia. J. Mater. Cycles Waste Manage. **8**(1), 1–12 (2006). https://doi.org/10.1007/s10163-005-0147-0
45. Wath, S.B., Dutt, P.S., Chakrabarti, T.: E-waste scenario in India, its management and implications. Environ. Monit. Assess. **172**(1–4), 249–262 (2011). https://doi.org/10.1007/s10661-010-1331-9
46. Widmer, R., Oswald-Krapf, H., Sinha-Khetriwal, D., Schnellmann, M., Böni, H.: Global perspectives on e-waste. Environ. Impact Assess. Rev. **25**(5 SPEC. ISS.), 436–458 (2005). https://doi.org/10.1016/j.eiar.2005.04.001
47. Gu, F., Ma, B., Guo, J., Summers, P.A., Hall, P.: Internet of things and big data as potential solutions to the problems in waste electrical and electronic equipment management: an exploratory study. Waste Manage. **68**, 434–448 (2017). https://doi.org/10.1016/j.wasman.2017.07.037
48. Kang, K.D., Kang, H., Ilankoon, I.M.S.K., Chong, C.Y.: Electronic waste collection systems using Internet of Things (IoT): household electronic waste management in Malaysia. J. Cleaner Prod. **252** (2020). https://doi.org/10.1016/j.jclepro.2019.119801
49. Maksimovic, M.: Leveraging internet of things to revolutionize waste management. Int. J. Agric. Environ. Inf. Syst. **9**(4), 1–13 (2018). https://doi.org/10.4018/IJAEIS.2018100101
50. Srikanth, C.S., Rayudu, T.B., Radhika, J., Anitha, R.: Smart waste management using internet-of-things (IoT). Int. J. Innovative Technol. Exploring Eng. **8**(9), 2518–2522 (2019). https://doi.org/10.35940/ijitee.g5334.078919
51. Giannopoulou, E., Mitrou, N.: An ai-based methodology for the automatic classification of a multiclass Ebook collection using information from the tables of contents. IEEE Access **8**, 218658–218675 (2020). https://doi.org/10.1109/ACCESS.2020.3041651

52. Gharibshah, Z., Zhu, X.: User response prediction in online advertising. ACM Comput Surv. Assoc. Comput. Mach. (2021). https://doi.org/10.1145/3446662
53. Sankar, H., Subramaniyaswamy, V., Vijayakumar, V., Arun Kumar, S., Logesh, R., Umamakeswari, A.: Intelligent sentiment analysis approach using edge computing-based deep learning technique. Softw.—Pract. Exper. **50**, 645–657. https://doi.org/10.1002/spe.2687
54. Altinisik, F., Yilmaz, H.H.: Predicting customers intending to cancel credit card subscriptions using machine learning algorithms: a case study. In: ELECO 2019—11th International Conference on Electrical and Electronics Engineering, pp. 916–920. Institute of Electrical and Electronics Engineers Inc. (2019). https://doi.org/10.23919/ELECO47770.2019.8990563
55. Shankar, P., Werner, N., Selinger, S., Janssen, O.: Artificial intelligence driven crop protection optimization for sustainable agriculture. In: IEEE/ITU International Conference on Artificial Intelligence for Good, AI4G 2020, pp. 1–6. Institute of Electrical and Electronics Engineers Inc. (2020) https://doi.org/10.1109/AI4G50087.2020.9311082
56. de Oliveira e Lucas, P., Alves, M.A., de Lima e Silva, P.C., Guimarães, F.G.: Reference evapotranspiration time series forecasting with ensemble of convolutional neural networks. Comput. Electron. Agric. **177** (2020).https://doi.org/10.1016/j.compag.2020.105700
57. Henderson, P., Hu, J., Romoff, J., Brunskill, E., Jurafsky, D., Pineau, J.: Towards the systematic reporting of the energy and carbon footprints of machine learning. J. Mach. Learn. Res. **21** (2020)
58. da Silva, I.N., Hernane Spatti, D., Andrade Flauzino, R., Liboni, L.H.B., dos Reis Alves, S.F.: Artificial neural network architectures and training processes. In: Artificial Neural Networks, pp. 21–28. (2016)https://doi.org/10.1007/978-3-319-43162-8_2
59. Omidvar, A., Pourmodheji, H., An, A., Edall, G.: Learning to determine the quality of news headlines. In: ICAART 2020—Proceedings of the 12th International Conference on Agents and Artificial Intelligence, vol. 1, pp. 401–409. SciTePress (2020). https://doi.org/10.5220/000 9367504010409
60. Eckart, L., Eckart, S., Enke, M.: A brief comparative study of the potentialities and limitations of machine-learning algorithms and statistical techniques. E3S Web Conf. **266**, 02001 (2021). https://doi.org/10.1051/e3sconf/202126602001
61. Kim, J., Lim, G., Kim, Y., Kim, B., Bae, C.: Deep learning algorithm using virtual environment data for self-driving car. In: International Conference on Artificial Intelligence in Information and Communication (ICAIIC) (2019). https://doi.org/10.1109/icaiic.2019.8669037
62. Roy, A., Sun, J., Mahoney, R., Alonzi, L., Adams, S., Beling, P.: Deep learning detecting fraud in credit card transactions. In: Systems and Information Engineering Design Symposium (SIEDS) (2018). https://doi.org/10.1109/sieds.2018.8374722
63. Reddy Karri, S.P., Santhosh Kumar, B.: Deep learning techniques for implementation of Chatbots. In: International Conference on Computer Communication and Informatics (ICCCI) (2020). https://doi.org/10.1109/iccci48352.2020.9104143
64. Pan, M., Liu, Y., Cao, J., Li, Y., Li, C., Chen, C.-H.: Visual recognition based on deep learning for navigation mark classification. IEEE Access **8**, 32767–32775 (2020). https://doi.org/10. 1109/access.2020.2973856
65. Bae, H.-S., Lee, H.-J., Lee, S.-G.: Voice recognition based on adaptive MFCC and deep learning. In: IEEE 11th Conference on Industrial Electronics and Applications (ICIEA). https://doi.org/ 10.1109/iciea.2016.7603830
66. Zhang, R., Isola, P., Efros, A.A.: Colorful image colorization. In: Computer Vision—ECCV, pp. 649–666 (2016). https://doi.org/10.1007/978-3-319-46487-9_40
67. Singh, S.P., Kumar, A., Darbari, H., Singh, L., Rastogi, A., Jain, S.: Machine translation using deep learning: an overview. In: International Conference on Computer, Communications and Electronics (Comptelix) (2017). https://doi.org/10.1109/comptelix.2017.8003957
68. Ali, H., Farman, H., Yar, H., Khan, Z., Habib, S., Ammar, A.: Deep learning-based election results prediction using Twitter activity. (2021). https://doi.org/10.21203/rs.3.rs-839553/v1
69. Chandra, S., Verma, S.: Big data and sustainable consumption: a review and research agenda. Vision (2021). https://doi.org/10.1177/09722629211022520

70. Chiu, A.S.F., Aviso, K.B., Baquillas, J., Tan, R.R.: Can disruptive events trigger transitions towards sustainable consumption? Cleaner Responsible Consumption **1**, 100001 (2020). https://doi.org/10.1016/j.clrc.2020.100001
71. Cenggoro, T.W., Wirastari, R.A., Rudianto, E., Mohadi, M.I., Ratj, D., Pardamean, B.: Deep learning as a vector embedding model for customer churn. In: Procedia Computer Science, vol. 179, pp. 624–631. Elsevier B.V. (2021). https://doi.org/10.1016/j.procs.2021.01.048
72. Kim, S., Choi, D., Lee, E., Rhee, W.: Churn prediction of mobile and online casual games using play log data. PLoS ONE **12**(7) (2017). https://doi.org/10.1371/journal.pone.0180735
73. De Vos, J.: The effect of COVID-19 and subsequent social distancing on travel behavior. Transp. Res. Interdisc. Perspect. **5**, 100121 (2020). https://doi.org/10.1016/j.trip.2020.100121
74. HT Auto DeskHT Auto Desk: Volvo announces subscribe-to-drive model for customers in Delhi. Hindustan Times Auto News (2021). https://auto.hindustantimes.com/auto/news/volvo-announces-subscribe-to-drive-model-for-indian-customers-41624342015927.html
75. Contreras-Castillo, J., Zeadally, S., Guerrero-Ibanez, J.A.: Internet of vehicles: architecture, protocols, and security. IEEE Internet Things J. **5**(5), 3701–3709 (2018). https://doi.org/10.1109/jiot.2017.2690902
76. Chen, H., Zhao, T., Li, C., Guo, Y.: Green internet of vehicles: architecture, enabling technologies, and applications. IEEE Access **7**, 179185–179198 (2019). https://doi.org/10.1109/ACCESS.2019.2958175
77. Udmale, P., Pal, I., Szabo, S., Pramanik, M., Large, A.: Global food security in the context of COVID-19: a scenario-based exploratory analysis. Prog. Disaster Sci. **7**, 100120 (2020). https://doi.org/10.1016/j.pdisas.2020.100120
78. de los Mozos, E.A., Badurdeen, F., Dossou, P.E.: Sustainable consumption by reducing food waste: a review of the current state and directions for future research. In: Procedia Manufacturing, vol. 51, pp. 1791–1798. Elsevier B.V (2020). https://doi.org/10.1016/j.promfg.2020.10.249
79. Food Waste Technology|Winnow. Winnowsolutions.com (2019). https://info.winnowsolutions.com/food-waste-technology
80. Asikis, T., Klinglmayr, J., Helbing, D., Pournaras, E.: How value-sensitive design can empower sustainable consumption. R. Soc. Open Sci. **8**(1) (2021). https://doi.org/10.1098/rsos.201418
81. Lenherr, N., Pawlitzek, R., Michel, B.: New universal sustainability metrics to assess edge intelligence. Sustain. Comput.: Inf. Syst. **31** (2021). https://doi.org/10.1016/j.suscom.2021.100580
82. Zhang, X., Manogaran, G., Muthu, B.A.: IoT enabled integrated system for green energy into smart cities. Sustain. Energy Technol. Assess. **46** (2021).https://doi.org/10.1016/j.seta.2021.101208
83. Ustundag Soykan, E., Bilgin, Z., Ersoy, M.A., Tomur, E.: Differentially private deep learning for load forecasting on smart grid. In: 2019 IEEE Globecom Workshops, GC Workshops 2019—Proceedings. Institute of Electrical and Electronics Engineers Inc. (2019). https://doi.org/10.1109/GCWkshps45667.2019.9024520
84. Kant, G., Sangwan, K.S.: Predictive modeling for power consumption in machining using artificial intelligence techniques. Procedia CIRP **26**, 403–407 (2015). https://doi.org/10.1016/j.procir.2014.07.072
85. Mehmood, H., Liao, D., Mahadeo, K.: A review of artificial intelligence applications to achieve water-related sustainable development goals. In: IEEE/ITU International Conference on Artificial Intelligence for Good (AI4G) (2020). https://doi.org/10.1109/ai4g50087.2020.9311018
86. Bejarano, G., Kulkarni, A., Raushan, R., Seetharam, A., Ramesh, A.: SWaP: probabilistic graphical and deep learning models for water consumption prediction. In: BuildSys 2019—Proceedings of the 6th ACM International Conference on Systems for Energy-Efficient Buildings, Cities, and Transportation, pp. 233–242. Association for Computing Machinery, Inc. (2019) https://doi.org/10.1145/3360322.3360846
87. Sanu, M.: 5 examples of how AI is helping companies become more sustainable. Winnowsolutions.com (2019). https://blog.winnowsolutions.com/5-examples-of-how-ai-is-helping-companies-become-more-sustainable

88. Prater, M.: 25 Google Search Statistics to Bookmark ASAP. Hubspot.com (2021). https://blog. hubspot.com/marketing/google-search-statistics
89. AI for Sustainable Water in Bengaluru: Indiaai.gov.in (2020). https://indiaai.gov.in/case-study/ ai-for-sustainable-water-in-bengaluru

Chapter 2
Deep Technologies Using Big Data in: Energy and Waste Management

Jyotsna Verma

Introduction

The relentless demands of urbanization and economic growth have resulted in the huge production of data, energy consumption, and wastes. The key and required element in technological change and urbanization is data, and with the evolution of handheld devices, data become the most prominent element in every domain. The production of global data has been increased exponentially due to the COVID-19 pandemic situation and is estimated that up to 2025 the generated data will rise by 180 zettabytes [1]. This huge production of data from a variety of sources, such as social, industrial, and transactional, demands high energy and environmental wastes. Also, the growth of global economy increased industrial activities, and limited natural resources lead to the energy crisis and environmental wastes. Particularly, the energy crisis is a big issue today, and the ultimate solution to this problem is the use of non-conventional energy sources and the optimized use of conventional energy. This solution helps contribute less environmental waste and can fulfill the demands of energy. Not only had the energy demand increased but, technological advancement also contributes lots of wastes. According to the European Union directive [2] "waste means any substance or object which the holder discards or intends or is required to discard." Humankind produces tons of global waste each year, and the proliferation of electrical and electronic equipment (EEE) worsens the situation. In 2019, 53.6 million metric tons of toxic wastes were reported from discarded EEE, and it is predicted that by 2030 global e-waste will reach 74 Mt [3]. The COVID-19 pandemic also added lots of personal protective equipment (PPE) wastes, and since the pandemic outbreak, it is estimated that worldwide 1.6 million tons of plastic

J. Verma (✉)
Department of Computer Science, The ICFAI University, Jaipur, India
e-mail: jyotsna.verma@iujaipur.edu.in

© The Author(s), under exclusive license to Springer Nature Singapore Pte Ltd. 2023
V. Kadyan et al. (eds.), *Deep Learning Technologies for the Sustainable Development Goals*, Advanced Technologies and Societal Change,
https://doi.org/10.1007/978-981-19-5723-9_2

waste are generated per day [4]. Hence, energy and waste management is necessary to reduce carbon footprints. To identify the non-conventional energy sources, to optimize the use of conventional energy and power, and to manage the environmental and EEE wastes, there is a requirement of analysis of data. Proper analysis of data at each step of data analytics helps in energy and waste management. Various studies and literature showcases that big data analytics helps in managing energy and wastes to a large extent. Data analytics is an important field for the development of commercial, public, and government sectors. It is a field of study where data are examined, cleansed, transformed, and modeled to discover useful information, and support decision-making. Earlier, the availability of small-scale data made data management and processing quite simple; still, for several real-world scenarios, the absence of sufficient and complete datasets makes the analysis of data difficult. But, the pervasive and colossal presence of data has added more responsibility for its processing and storage. Hence, keeping in view of today's real-world scenarios, managing, analyzing, and storing enormous data with the traditional methods and processing tools is not possible, and there is a requirement of big data technology.

Big data are the potential solution to the problem, and in near future, it will find its way into every domain. The applications of big data analytics in diverse sectors, such as health care, industries, businesses, energy management, waste management, bioinformatics, made big data analytics more promising and hot topic today. Big data are a field that deals with the extremely large volume of data that is analyzed to find patterns and to have valuable insight into the trends and prospects of markets and real-world applications. The key characteristic of big data is what most of us think is handling extremely large datasets, but, till now, it is not defined that presence of what amount of data comes under the big data environment. It is not the amount of data that matters, more than the amount of data focus of big data is on the analytics and what to do with those datasets. So, combining big data, with high analytics, helps to get benefits in various application domains with reduced processing time, costs, optimized processes, and smart decisions. There are different types of analytics available for the analysis of heterogeneous datasets: (1) Descriptive analytics: Descriptive analytics is the simplest data analytics that analyzes real-time or historical data. The past data are analyzed with the statistical tools and techniques for the perception of future potentialities and to have an insight of what went wrong or proper to get a clear view of outcomes. (2) Predictive analytics: It analyzes the past data patterns and trends to predict the possibilities of future outcomes by using statistical and machine learning algorithms. (3) Prescriptive analytics: The perspective analysis is the next step after predictive analysis which analyzes the future outcome and provides the number of actions that need to be taken based on the outcomes. (4) Diagnostic analytics: The detailed analysis of data is done in diagnostic analysis and requires enough data to identify anomalies and relationships between the data.

Big data analytics provides a variety of opportunities to develop novel models for the management of energy and wastes. Still, high-dimensional real-world problem, uncertain and incomplete datasets, inaccessible data, and poor quality data add lots of complexity in designing models for energy and waste management. To deal with these issues, researchers are going in direction of deep learning techniques which can be used with big data analytics for energy and waste management. Deep learning is a subfield of machine learning which uses statistical machine learning techniques for learning representation of data based on artificial neural networks simulating neurons of the human brain. Deep learning techniques make the data collection, analysis, and interpretation of data faster and easy. Hence, enabling deep learning techniques with big data technology helps in solving complex problems, dealing with incomplete and uncertain datasets efficiently. Dealing with the energy crisis and waste management is not an easy task. Several challenges need to be dealt like bringing reforms in government policies, identifying waste sources, optimizing conventional energies, and accurate and sufficient data collection. Exploiting deep learning techniques in combination with big data analytics provides aid in solving these issues and helps in maintaining a sustainable environment.

Deep Learning in Big Data Analytics

The concept of big data is not new; it has been in use since the 1990s. However, the concept gained wide popularity in the last two decades because of the variety, volume, and velocity of data and has made pavement for various data analysis techniques available today. The real challenge of data analytics in today's scenario is the analysis of the huge amount of data. The data can be in any form and is generating with great velocity. Basically, big data are a collection of structured, unstructured, and semi-structured data. Structured data are well defined, highly organized, and quantitative type of data like names, addresses, dates, credit card numbers, geolocations, and data from sensors and network servers, which is usually stored in data warehouses. These data are easy to store, analyze, and process using the predefined or fixed format. Unstructured data are an unorganized and qualitative type of data like social media activity, surveillance data, images, audio, weather data, records emails, sensor data, and rich media, which are usually stored in the data lake. The volume, velocity, variety have made big data analytics difficult and require more understanding and processing capabilities. Moreover, real-world problem space with high-dimensional and uncertain data adds complexity to the data analysis process. Therefore, there is a requirement of efficient strategies, algorithms, and techniques for big data analytics that run through different datasets in order to draw conclusions. For that purpose, many researchers have proposed various data analysis techniques by using deep learning algorithms for handling different problems of huge datasets [5–9]. The variety and low-quality data [10–12] generated from the digital data explosion pose other issues in big data analytics. Deep learning models were designed for handling heterogeneous data representation [13–18] and low-quality data [19–21] which gives strong evidence of efficient handling of data analytics problems by the deep learning algorithms. Also, a lot of works on designing the processor architecture and data layouts are being done to improve the computing power for speeding up the learning process, and it proved enormous capability in big data deep learning [22–27].

Deep learning is an astounding machine learning technique that uses supervised unsupervised, and reinforcement learning techniques to learn multilevel representation and feature extraction from huge datasets. It intricate relationships between a large number of interdependent variables for the accurate pattern recognition and classification of data [28–31]. Deep learning algorithms are efficient in solving complex and computationally expensive problems, and hence, they can handle the challenges of big data analytics problems very well. The focus of the chapter is how big data analytics exploits deep learning techniques to provide solutions for energy and waste management; hence, the chapter will not extensively cover the concept of deep learning and big data analytics, but only present a brief overview of it.

Exploiting Deep Learning for Energy Management in Big Data

Over time, escalation in demand for handheld devices and economic growth have increased energy consumption. Even though renewable energy is growing at a rapid rate around the world, fossil fuels remain to account for the majority of total energy consumption. But, in 2020, there was a decline in the global energy consumption growth by 4% due to COVID-19 pandemic situation. The energy demand is prevalent in every sector, but the industry sector is the largest energy consumer. With the depletion of fossil fuels, it is necessary to reduce energy use and provide incentives to non-conventional energy sources, such as solar, wind, bio, hydro, and tidal energy. The efficient management of energy ensures optimum use of energy resources and reduces energy costs [32–36]. The development of energy management programs [37–39] proved to save energy, cost, reduces energy wastes and emission of greenhouse gases. However, traditional methods of energy management face several difficulties, including operational efficiency, environmental wastes, resilience, and stability of energy infrastructure. Hence, there is a necessity for big data analytics to cope with these difficulties and establish an efficient, green, and cost-effective energy management system.

Introduction

The way energy is produced and consumed is changing with the inception of big data analytics. Several smart energy systems with big data analytics are available in the literature. The smart energy system is actually a combination of various technologies, sensors, and intelligent devices, and hence, it works with heterogeneous data coming from different domains. The energy production, weather data, thermostat data, data of electric car, power supply, clean power, real state, energy databases, behavioral pricing, and behavioral analytics to control the energy consumption are considered 10

methods through which big data can reshape the energy [40]. The geographical information system is also a conventional data source that provides geographic attributes,
features, environment, and resources of a particular geographical region. GIS does not
collect real-time data; it takes time to update the recorded data and helps the energy
system as effective decision support [41–44]. The heterogeneity of data sources and
data in different formats makes database integration an important part of the energy
system in big data analytics [45–49]. Not much work has been done for the integration
of complex data coming from the enterprise, energy and production systems [50].
Researchers are usually focused on the development of energy management models,
but complex and heterogeneous datasets must be integrated with the energy management systems in order to support an effective decision-making process. Yet, high
dimensionalities, uncertainty, and incompleteness in the datasets pose difficulties in
the decision-making process of energy management. Hence, deep learning techniques
are required for the analysis of datasets with unreasonable complexities to harvest
useful information, patterns, etc., to predict and optimize the energy consumption
requirements. The deep learning-enabled energy system helps in finding solutions to
various big data analytics problems pertaining to energy systems in every domain,
such as load classification problems in smart grid, energy demand prediction, high-
performance computing problems, data integration problems, and optimum use of
renewable energy sources.

Analytics Model for Energy Management Using Deep Learning

To fully utilize the potential of big data analytics for energy management, there is a
requirement of a process model. The process model is a basic systematic framework
for smart energy management. The first step in the model is the collection and storage
of the data gathered from the different sources. The stored data are then cleaned and
processed which transform the dataset into organized and structured datasets. The
structured datasets are then integrated, and relevant variables are selected from the
integrated datasets for further processing. The processed datasets are then graphically
represented with the data visualization methods to discover patterns and relationships
among variables in the datasets. Based on discovered patterns, an intelligent decision
will be made with the real-time interaction for smart energy management. For the
efficient energy system, there is a requirement of applying different analytics at each
phase of the process model. Thus, various deep learning techniques are amalgamated
with the data analytics energy model to provide a better decision-making support
system.

Predictive Analysis

Predictive analysis is one such data analysis method that uses a historical dataset for the prediction of future energy consumption possibilities. Literature discloses a predictive analysis energy model [51] which was based on the deep neural network technique for the development of the smart city. Another predictive data analysis on the small factory with the historical data was done for the smart energy management in the manufacturing unit [52]. Moreover, a predictive analysis energy model using supervised learning for high-performance computing was also developed [53].

Descriptive and Time Series Analysis

The descriptive analysis deals with past and present datasets to predict the outcome whereas time series analytics deals with the series of observations of specific or numerical variables listed with the date or timestamp. Both the analytics method is the part of predictive analytics. With the time series analytics, an energy model based on deep learning techniques was developed which analyzes the pattern of nighttime light usage and Twitter data volume for estimating the domestic electric consumption [54]. A time series dataset of household power consumption was used for the analysis of energy demand using the deep learning technique for the minimization of energy consumption [55]. Also, a deep learning-based fuel consumption model was developed which chooses the optimum sea route for the container ships [56].

Exploratory Analysis

For the purpose of increasing the energy efficiency of residential buildings, an exploratory analysis was done by analyzing the cross-country energy consumption patterns in the European Union [57]. In other real-world scenario, the infrastructure design, traffic conditions, driving behavior of electric vehicles in Beijing were analyzed to improve the energy efficiency at different traffic conditions and infrastructural design [58].

Prescriptive Analysis

The prescriptive analysis provides the optimal actions based on the outcomes of the data analysis process. A study was done with the Zenodo dataset which includes the social, economic, behavioral, and geographic factors [59]. The study analyzes the energy-related behavior by assessing the social and economic behavior of occupants at tertiary buildings. Moreover, the outcome of the analytical study identifies the number of greener actions needs to take at the workplace and barriers to unnecessary energy consumptions. Another study on the daylight performance of the commercial buildings in Egypt was conducted [60]. A prescriptive analysis coupled with performance guidelines was done to improve energy efficiency.

Discussion

Big data analytics with deep learning techniques actually minimizes energy consumption. The data analytic energy models help to predict the energy demand based on the analysis of demographic and environmental needs. The prediction of energy demand, load balancing, and classification usually depends upon the scale of data. Short-term dataset usually contains daily or hourly data whereas long-term dataset contains yearly data. Small-scale data cannot represent the whole sample data, and large-scale datasets require lots of computational effort and cost for the representation of a particular sample space. Usually, data analytics studies are done on one month to one-year-old dataset records [61]. Long-term record sets are not usually used for the data analysis because a lot of uncertainties are involved with the datasets [61] and require long-term energy prediction models [62]. Apart from the dataset granularity, the energy consumption model for commercial, educational, and residential buildings also plays a vital role in energy management. The residential building consumes more energy, yet researchers are more focused on developing energy consumption models for commercial and non-residential buildings [61]. The absence of sensor data collection infrastructure in residential buildings is the main reason behind the fewer prediction models for residential buildings.

Deep learning algorithms are required to train the data analytics energy model which in turn offers optimized solutions. Several energy models and systems are developed to improve energy consumption, efficiency, and specific energy requirements in all sectors. In smart grid technology, deep learning offers a more decentralized and efficient energy model with high operational intelligence and decision-making capabilities. Deep learning deals with load forecasting [63], load classification [64], and energy demand problems in smart grids very well [65–68]. The dynamic pricing [69] also helps in saving energy; efficient methods and strategies are developed depending upon the customer energy consumption behavior analysis for the identification of grid loss and energy diversion [70–74]. Of the entire sector, the industrial sector consumes more energy. A study was conducted on manufacturing industries with two deep neural network techniques to predict the building energy profiles [75]. The forecast helps in the identification of energy consumption, environmental conditions, undesirable and significant costs involved in the manufacturing sector. Other than the industrial sector, energy models for transportation and commercial sectors are also developed based on deep learning techniques [76]. The increased need for power and the desire to reduce carbon emissions paved the way for the usages of renewable energy, such as wind power, thermal energy, and solar energy. But, the inherent intermittency of the bulk of renewable energy supplies is making it difficult to enhance their penetration into the energy mix. Literature reveals various effective renewable energy models based on deep learning techniques with big datasets. A deep neural network-based wind power prediction system on unseen data was developed which reduces training time on wind farm datasets [77]. Another deep learning-based prediction model for wind power forecasting was developed with the Maine wind farm ISO NE, USA dataset [78] to maintain demand and supply of energy in an optimum way. Likewise, thermal and solar energy production forecast through deep learning approaches is also done to improve the usage of renewable energies [79–84]. Considering the above, big data analytics along with deep learning techniques provides efficient solutions for several problems of energy management systems.

Exploiting Deep Learning for Waste Management in Big Data

The European Union directive on waste [85] is concerned and repeals certain directives in order to take *"measures to protect the environment and human health by preventing or reducing the adverse impacts of the generation and management of waste and by reducing overall impacts of resource use and improving the efficiency of such use."* Recycling waste materials and goods in times of resource scarcity are not new; generally, to cover, society needs mankind to fix greater economic and societal value to the wastes. There are various sources of waste, and it can be divided into the following categories based on the source:

1. City waste: Household, demolition, and commercial wastes are all included in city waste.
2. Hazardous wastes: This category comprises electronic waste, industrial waste, radioactive trash, and explosive garbage.
3. Biomedical wastes: Clinical wastes are included in the category of biomedical wastes.

The technological advances resulted in an increase in wastes. According to the United Nations report 2019, 44 million tons of e-waste are produced each year, and substantially, only, 20% are recycled. Only, India has contributed 3.2 million tons of e-waste last year. The 2020 global e-waste report shows that globally 53.6 million tons of e-waste were produced in 2019, and it will go up by 20% in the next five years. With the global COVID-19 Pandemic, India alone is producing 146 tons of biomedical waste every day as reported in the central pollution control board. Of all biomedical wastes, 85% of wastes are non-hazardous, and 15% of the wastes are infectious, hazardous, and toxic. Apart from biomedical wastes, a large portion of solid wastes is generated from industries. The traditional methods of managing wastes are costly [86] which made industries reluctant toward adopting proper waste management methods [87, 88]. The absence of strict regulations and insufficient disposal policies has encouraged the industries, municipal corporations, and hospitals to manage wastes in their own arbitrary manner [88, 89]. Following the current growth rate of hazardous or non-hazardous wastes, it is estimated to increase millions of tons in coming years which needs proper attention, preventive measures, and strict disposal policy else it hampers the environment and health of mankind.

Introduction

Global waste is generated on large scale, and to manage these wastes, there is a requirement of waste intelligence methods. The waste intelligence method manages the waste efficiently as compared to conventional methods. It helps in classification, sorting, identification of the source of waste, and amount of global waste production. Big data analytics and other enabling technologies like IoT, deep learning techniques help in analyzing wastes production and reduction by providing better data analytics techniques and models which in advance predict the amount of waste generation. The prediction helps in proper recycling and minimizing the wastes. In the waste management process, the most preferred and initial step is the reduction of wastes. The most effective way to manage waste is by reducing industry waste, city, and biomedical wastes. The second step is to reuse or recycle the wastes. Reuse and recycling involve exploratory analysis. The least preferred step is landfilling and dumping. The landfilling involves prescriptive analysis and hampers the environment very adversely. The absence of awareness in society makes the segregation of wastes difficult which in turn affect the environment adversely. But, in reality, landfilling and dumping are a widely used method to manage the wastes which are making a bad impact on the environment. The use of big data analytics with deep learning methods will improve the situation and make reuse and recycling a better and easy option for managing the wastes.

Analysis Model of Waste Management Using Deep Learning

For proper management of wastes, the analysis of wastes is important and hence required an efficient analysis model. Big data analytics is required at each step of the waste management process. Figure 2.1 shows the process of waste management analysis. In the waste management analysis process, the first step is to get the required data. The required data can be collected through questioners, interviews, or by deploying sensor networks in an application domain. In particular to wastes, the details of waste can be gathered through municipal corporations; in case of city wastes, the biomedical wastes datasets can be gathered from the hospitals, clinics, etc., and industrial wastes can be gathered from the industries. The collected data are then preprocessed for analysis. The dataset can be unstructured or incomplete, so structuring of data is important as required for the analysis model, statistical tools, etc. The structured data can contain errors, duplicity, or it can be incomplete, so

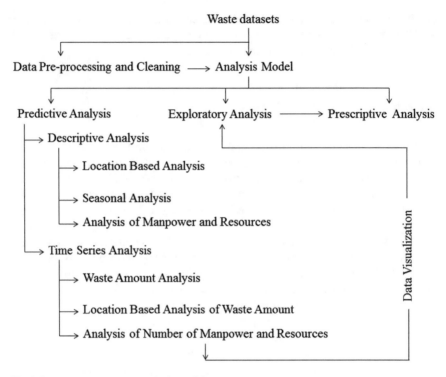

Fig. 2.1 Waste management analysis model process

data cleaning is also important before the analysis of the datasets for getting accurate information to forecast the trends and prospects. The next step is to analyze the processed datasets, and hence, datasets are then applied to the appropriate analysis models for managing the wastes.

Predictive Analysis

The cleaned dataset is first fed to the predictive analysis model. The predictive analysis uses the waste datasets for predicting future outcomes with the help of deep learning and statistical techniques. In this analysis model, sources of wastes are identified for reducing the amount of waste. A predictive analysis model associated with big data [90] was developed using R analytical tool for forecasting the number of possibilities in advance. This is the preliminary step which uses the descriptive analysis phase and time series analysis phase.

(i)   **Descriptive analysis**: In this phase, the datasets are analyzed to identify the location of wastes, tracking the seasonal waste patterns, manpower (cleaner, garbage collector, rag pickers, etc.), and other resources required. An image recognition system was developed [91] using deep learning convolutional neural network for the identification and classification of e-wastes from the photos uploaded by the individual to the waste collection company's server. The waste collection company then collects the e-waste according to the need of manpower and the location. A quantum GIS-based descriptive and predictive analysis approach [92] was used for the production of in-time waste information. The real dataset of Jeju Island, South Korea, was used for the evaluation of the approach, and the findings revealed that predictive analytic models are better suited for waste management operations optimization and planning. Another multitask learning model based on a convolutional neural network was proposed for the localization and recognition of wastes [93].

(ii) **Time series analysis**: In this phase, the amount of waste generated is analyzed; analysis of waste on the basis of location is done, and the number of manpower and other resources required is identified. A method was designed for the minimization of food waste in the retail sector using an object detection algorithm and time series analysis [94]. The method analyzes the food wastes thrown in the smart bins. The analysis helps to optimize the raw materials used in the preparation of food items and thus minimizes the wastes. Another model was designed using deep learning techniques with the time series data for solid waste management [95].

The descriptive and time series analysis of datasets is then visualized. The data visualization represents the data in the form of graphs, plots, maps, charts, or other visual representations in order to predict and understand the patterns, trends, and relationships among datasets. Data visualization is very important when we are dealing with a massive amount of data. The visualized data are then used for exploratory analysis.

Exploratory Analysis

The exploratory analysis uses the visual representations generated from data visualization methods to discover patterns and understanding the main characteristics of the datasets. The exploratory analysis is beyond the formal hypothesis and data modeling testing processes. It gives a better understanding and insight into the relationship among the variables of the dataset. It helps in identifying the area where waste can be reused and applications for recycling the wastes. Various statistical and deep learning techniques are applied for the exploratory analysis of datasets. A classification model [96] is proposed based on deep learning for the classification of recyclable waste images. The TrashNet dataset was used for the classification of recyclable wastes and shows 95.87% classification accuracy when compared with other deep learning classification models. Another intelligent solid waste classification model [97] was proposed using convolutional neural networks which classify wastes according to their content like glass, paper, plastic, and organic wastes.

Prescriptive Analysis

On the basis of results generated from the exploratory analysis, the prescriptive analysis finds the number of actions that need to be taken to dispose of the wastes properly. The landfilling is mostly done through prescriptive analysis. It optimizes the results generated from the exploratory analysis and predicts the best possible action for disposing of the wastes. A multilayer hybrid deep learning system [98] was designed for sorting disposed wastes. The system deploys cameras and sensors at different places to gather the images of wastes which automatically classify the recyclable wastes and other things. A study was conducted in Italy [99] with cluster analysis and exploratory spatial data analytics techniques for the identification of spatial clusters and correlation in the distribution of landfill disposal. The analysis helped in framing appropriate guidelines for the actions which need to be taken for improving waste management. The prescriptive performance-based approach is fundamental for the promotion of less polluting practices. The proposed approach [100] analyzes the operational discharge system to the marine environment of Norway.

Discussion

Practically, monitoring and managing wastes are very difficult. Traditional waste management techniques are incapable of reducing, reusing, and recycling waste properly. So, embedding wastes with sensors and big data analytics is very important as it gives detailed insights and real-time monitoring. But, real-time monitoring generates dynamic data of high dimensionality, also generates uncertain and incomplete data. Hence, to address those problems, researchers are applying deep learning techniques on various phases of the analytics process as well as helps in designing various waste management systems and are providing efficient results. One such smart waste management system was developed using LoRa communication protocol and TensorFlow-based deep learning model [101]. In this system, the sensor data about the location and filling level of the bin were sent through the LoRa protocol, and the detection and classification of objects were done through TensorFlow with a pre-trained object detection model. The limited availability of materials for the identification, separation of solid wastes poses another challenge. To address these issues, a method [102] was developed which recognizes the material by training a deep convolutional network. For the training, a database of images was created with different combinations of cameras, data augmentation techniques, and illuminations. Households also play a major factor in the production of wastes. A multi-site long short-term memory (LSTM) neural network was developed which forecasts the rate at which the waste is generated from the households [103]. The proposed model was designed for the households of Herning, Denmark and uses the historical data of their weekly waste weights. In order to create sustainable and cost-effective waste management operations, data on waste generation are collected extensively. In addition, the wide availability of geospatial data from the different registries of application domains has made researchers design novel data analytics approaches for more detailed and in-time waste generation information [104]. Various data-based approaches have been designed by researchers for the optimization of waste collection and recycling [104, 105].

Challenges and Prospects: Energy and Waste Management

Energy and waste management in big data analytics deals with various issues and challenges which need to be addressed properly. Following are the issues or problems of energy and waste management in big data analytics using deep learning techniques:

1. **Data granularity**: Deep learning modeling with the small-scale data is difficult as it requires a large amount of data for training and learning. The generation of datasets for energy and waste management on large scale for all the applications is difficult and requires huge computational effort and costs. Training deep learning models with small-scale data will lead to overfitting problems [106]. However, data augmentation attempts to solve this problem to a great extent [107].
2. **Integrated systems**: Energy and waste management is not an easy task; it requires big data analytics with various enabling techniques for modeling energy and waste systems. Cloud computing, deep learning algorithms, the Internet of Things, and many more other technologies can be integrated with big data analytics to provide efficient energy and waste management system.
3. **Government policies and regulations**: There is a substantial need for regulations and policies in the industries and commercial sectors regarding energy management and waste disposal for preserving the environment.
4. **Sustainable environment system**: To solve the energy crisis and to maintain a sustainable environment, waste to energy plants should be constructed. The waste to energy plants can generate lots of energy and can help the industrial sector in a big way. Big data analytics with deep learning algorithms can help to identify the requirement of resources, skilled manpower, energy demand, etc.
5. **Data integration**: Heterogeneous data standards and protocols have been adopted by energy and waste management organizations. Due to this, data integration from various sources is a big problem. Various techniques and models have been proposed to solve the problems of data integration. Still enabling big data analytics with other technologies poses challenges in data integration and data sharing.
6. **Data processing and analysis**: Several energy models and waste systems have been developed by using deep learning algorithms with big data analytics. Still, these developed models face limitations of data granularity, presence of spatial and temporal datasets which pose barriers in the processing and analysis of data.
7. **Privacy and security**: Maintaining the security and privacy of datasets is very important. The energy system involves a lot of security concerns and is vulnerable

to be attacked. The data from commercial and residential sectors about electricity usage and demand need to be protected from attackers as it involves personal data.

Summary

The recent technological advances and escalation in big data analytics gradually lead to the energy crisis and production of massive environmental wastes. The energy crisis is a bottleneck problem because of the limited natural resources and absence of an efficient energy model for various application scenarios. Ensembling big data analytics with deep learning algorithms provides solutions for energy demand and crisis. Deep learning offers the optimized solution and helps in the identification of energy demand and consumption by the individual, commercial, and industrial sectors. The energy crisis and global wastes are major concerns in today's scenario, especially after the COVID-19 pandemic, there has been a spike in biomedical wastes and is creating lots of environmental problems. With the limited resources, varied content, dynamic and fast generation of data, it is impossible to manage energy crisis and wastes with the traditional methods. As the processing of wastes with conventional methods is expensive, the industries and municipal corporations are managing their waste arbitrarily by taking advantage of the absence of awareness in society and loosely constructed policies and regulations. Hence, there is a need for strict regulations and deep learning techniques and tools which will help in solving these computationally expensive problems. Big data analytics along with deep learning technologies will give an accurate insight into datasets patterns which helps in managing energy and wastes. This chapter discusses the energy and waste management issues with deep learning techniques in big data analytics. Moreover, the prospects of energy and waste management using big data analytics and deep learning techniques are also presented.

References

1. https://www.statista.com/statistics/871513/worldwide-data-created/. Last accessed on 2021/7/3
2. Directive, E.C.: Directive 2008/98/EC of the European Parliament and of the Council of 19 November 2008 on waste and repealing certain Directives. Official J. Eur. Union L **312**(3) (2008)
3. https://www.itu.int/en/ITU-D/Environment/Pages/Spotlight/Global-Ewaste-Monitor-2020.aspx
4. Benson, N.U., Bassey, D.E., Palanisami, T.: COVID pollution: impact of COVID-19 pandemic on global plastic waste footprint. Heliyon **7**(2), e06343 (2021)
5. Raina, R., Madhavan, A., Ng, A.Y.: Large-scale deep unsupervised learning using graphics processors. In Proceedings of the 26th Annual International Conference on Machine Learning, pp. 873–880 (2009)
6. Cireşan, D.C., Meier, U., Gambardella, L.M., Schmidhuber, J.: Deep, big, simple neural nets for handwritten digit recognition. Neural Comput. **22**(12), 3207–3220 (2010)

7. Dahl, G.E., Yu, D., Deng, L., Acero, A.: Context-dependent pre-trained deep neural networks for large-vocabulary speech recognition. IEEE Trans. Audio Speech Lang. Process. **20**(1), 30–42 (2011)
8. Hinton, G., Deng, L., Yu, D., Dahl, G.E., Mohamed, A.R., Jaitly, N., et al.: Deep neural networks for acoustic modeling in speech recognition: the shared views of four research groups. IEEE Signal Process. Mag. **29**(6), 82–97 (2012)
9. Deng, L., Yu, D., Platt, J.: Scalable stacking and learning for building deep architectures. In 2012 IEEE International Conference on Acoustics, Speech and Signal Processing (ICASSP), pp. 2133–2136. IEEE (2012)
10. Saha, B., Srivastava, D.: Data quality: the other face of big data. In: IEEE 30th International Conference on Data Engineering, pp. 1294–1297. IEEE (2014)
11. Becker, D., King, T.D., McMullen, B.: Big data, big data quality problem. In: IEEE International Conference on Big Data (Big Data), pp. 2644–2653. IEEE (2015)
12. Li, W., Xu, S., Peng, X.: Research on comprehensive evaluation of data source quality in big data environment. Int. J. Comput. Intell. Syst. (2021)
13. Srivastava, N., Salakhutdinov, R.: Multimodal learning with deep Boltzmann machines. In: NIPS, vol. 1, p. 2 (2012)
14. Ouyang, W., Chu, X., Wang, X.: Multi-source deep learning for human pose estimation. In: Proceedings of the IEEE Conference on Computer Vision and Pattern Recognition, pp. 2329–2336 (2014)
15. Zhao, L., Hu, Q., Wang, W.: Heterogeneous feature selection with multi-modal deep neural networks and sparse group LASSO. IEEE Trans. Multimedia **17**(11), 1936–1948 (2015)
16. Venugopalan, J., Tong, L., Hassanzadeh, H.R., Wang, M.D.: Multimodal deep learning models for early detection of Alzheimer's disease stage. Sci. Rep. **11**(1), 1–13 (2021)
17. Liu, D., Chen, L., Wang, Z., Diao, G.: Speech expression multimodal emotion recognition based on deep belief network. J. Grid Comput. **19**(2), 1–13 (2021)
18. Li, H., Huang, J., Huang, J., Chai, S., Zhao, L., Xia, Y.: Deep multimodal learning and fusion based intelligent fault diagnosis approach. J. Beijing Inst. Technol. **30**(2), 172–185 (2021)
19. Vincent, P., Larochelle, H., Bengio, Y., Manzagol, P.A.: Extracting and composing robust features with denoising autoencoders. In: Proceedings of the 25th International Conference on Machine Learning, pp. 1096–1103 (2008)
20. Vincent, P., Larochelle, H., Lajoie, I., Bengio, Y., Manzagol, P.A., Bottou, L.: Stacked denoising autoencoders: learning useful representations in a deep network with a local denoising criterion. J. Mach. Learn. Res. **11**(12) (2010)
21. Li, R., Gao, H.: Denoising and feature extraction of weld seam profiles by stacked denoising autoencoder. Weld World 1–9 (2021)
22. Vanhoucke, V., Senior, A., Mao, M.Z.: Improving the speed of neural networks on CPUs (2011)
23. Krizhevsky, A., Hinton, G.: Learning multiple layers of features from tiny images (2009)
24. Farabet, C., LeCun, Y., Kavukcuoglu, K., Culurciello, E., Martini, B., Akselrod, P., Talay, S.: Large-scale FPGA-based convolutional networks. In: Scaling up Machine Learning: Parallel and Distributed Approaches, vol. 13, no. 3, pp. 399–419 (2011)
25. Qiu, C., Wang, X.A., Zhao, T., Li, Q., Wang, B., Wang, H.: An FPGA-based convolutional neural network coprocessor. Wireless Commun. Mobile Comput. **2021** (2021)
26. Yoo, J., Lee, D., Son, C., Jung, S., Yoo, B., Choi, C., et al.: RaScaNet: learning tiny models by raster-scanning images. In: Proceedings of the IEEE/CVF Conference on Computer Vision and Pattern Recognition, pp. 13673–13682 (2021)
27. Shehzad, F., Rashid, M., Sinky, M.H., Alotaibi, S.S., Zia, M.Y.I.: A scalable system-on-chip acceleration for deep neural networks. IEEE Access (2021)
28. Wang, J., Chen, Y., Hao, S., Peng, X., Hu, L.: Deep learning for sensor-based activity recognition: a survey. Pattern Recogn. Lett. **119**, 3–11 (2019)
29. Sun, Y., Wang, X., Tang, X.: Deep learning face representation from predicting 10,000 classes. In: Proceedings of the IEEE Conference on Computer Vision and Pattern Recognition, pp 1891–1898 (2014)

30. Hassan, A., Mahmood, A.: Convolutional recurrent deep learning model for sentence classification. IEEE Access **6**, 13949–13957 (2018)
31. Lakshmanaprabu, S.K., Mohanty, S.N., Shankar, K., Arunkumar, N., Ramirez, G.: Optimal deep learning model for classification of lung cancer on CT images. Futur. Gener. Comput. Syst. **92**, 374–382 (2019)
32. Pye, M., McKane, A.: Making a stronger case for industrial energy efficiency by quantifying non-energy benefits. Resour. Conserv. Recycl. **28**(3–4), 171–183 (2000)
33. Bunse, K., Vodicka, M., Schönsleben, P., Brülhart, M., Ernst, F.O.: Integrating energy efficiency performance in production management–gap analysis between industrial needs and scientific literature. J. Clean. Prod. **19**(6–7), 667–679 (2011)
34. Conticelli, E., Proli, S., Tondelli, S.: Integrating energy efficiency and urban densification policies: two Italian case studies. Energy Build. **155**, 308–323 (2017)
35. Cagno, E., Neri, A., Trianni, A.: Broadening to sustainability the perspective of industrial decision-makers on the energy efficiency measures adoption: some empirical evidence. Energ. Effi. **11**(5), 1193–1210 (2018)
36. Bin Abdulwahed, F.F.A.: The hidden benefits of energy efficiency: quantifying the impact of non-energy benefits when energy efficiency measures are implemented in the EU Iron and steel industry (Master's thesis) (2021)
37. Introna, V., Cesarotti, V., Benedetti, M., Biagiotti, S., Rotunno, R.: Energy management maturity model: an organizational tool to foster the continuous reduction of energy consumption in companies. J. Clean. Prod. **83**, 108–117 (2014)
38. Antunes, P., Carreira, P., da Silva, M.M.: Towards an energy management maturity model. Energy Policy **73**, 803–814 (2014)
39. Sola, A.V., Mota, C.M.: Influencing factors on energy management in industries. J. Clean. Prod. **248**, 119263 (2020)
40. GigaOM, 10 ways big data is remaking energy. https://gigaom.com/2012/01/29/10-ways-big-data-is-remaking-energy/
41. Voivontas, D., Assimacopoulos, D., Mourelatos, A., Corominas, J.: Evaluation of renewable energy potential using a GIS decision support system. Renewable Energy **13**(3), 333–344 (1998)
42. Jakubiec, J.A., Reinhart, C.F.: A method for predicting city-wide electricity gains from photovoltaic panels based on LiDAR and GIS data combined with hourly Daysim simulations. Sol. Energy **93**, 127–143 (2013)
43. Feng, J., Feng, L., Wang, J., King, C.W.: Evaluation of the onshore wind energy potential in mainland China—based on GIS modeling and EROI analysis. Resour. Conserv. Recycl. **152**, 104484 (2020)
44. Ferla, G., Caputo, P., Colaninno, N., Morello, E.: Urban greenery management and energy planning: a GIS-based potential evaluation of pruning by-products for energy application for the city of Milan. Renewable Energy **160**, 185–195 (2020)
45. Parent, C., Spaccapietra, S.: Issues and approaches of database integration. Commun. ACM **41**(5es), 166–178 (1998)
46. Devogele, T., Parent, C., Spaccapietra, S.: On spatial database integration. Int. J. Geogr. Inf. Sci. **12**(4), 335–352 (1998)
47. Stencel, K.: A data model for heterogeneous data integration architecture
48. Hasan, F.F., Bakar, M.S.A.: An approach for data transformation in homogeneous and heterogeneous information systems. In: 3rd International Congress on Human-Computer Interaction, Optimization and Robotic Applications (HORA), pp. 1–5. IEEE (2021)
49. Mhammedi, S., Gherabi, N.: Heterogeneous integration of big data using semantic web technologies. In: Intelligent Systems in Big Data, Semantic Web and Machine Learning, pp. 167–177. Springer, Cham (2021)
50. Bevilacqua, M., Ciarapica, F.E., Diamantini, C., Potena, D.: Big data analytics methodologies applied at energy management in industrial sector: a case study. Int. J. RF Technol. **8**(3), 105–122 (2017)

51. Tian, Y., Yu, J., Zhao, A.: Predictive model of energy consumption for office building by using improved GWO-BP. Energy Rep. **6**, 620–627 (2020)
52. Kumar, M., Shenbagaraman, V.M., Shaw, R.N., Ghosh, A.: Predictive data analysis for energy management of a smart factory leading to sustainability. In: Innovations in Electrical and Electronic Engineering, pp. 765–773. Springer, Singapore (2021)
53. Ozer, G., Garg, S., Davoudi, N., Poerwawinata, G., Maiterth, M., Netti, A., Tafani, D.: Towards a predictive energy model for HPC runtime systems using supervised learning. In: European Conference on Parallel Processing, pp. 626–638. Springer, Cham (2019)
54. Thonipara, A., Runst, P., Ochsner, C., Bizer, K.: Energy efficiency of residential buildings in the European Union—an exploratory analysis of cross-country consumption patterns. Energy Policy **129**, 1156–1167 (2019)
55. Sachin, M.M., Baby, M.P., Ponraj, A.S.: Analysis of energy consumption using RNN-LSTM and ARIMA Model. In: J. Phys.: Conf. Ser. **1716**(1), 012048 (2020)
56. Bui-Duy, L., Vu-Thi-Minh, N.: Utilization of a deep learning-based fuel consumption model in choosing a liner shipping route for container ships in Asia. Asian J. Shipping Logistics **37**(1), 1–11 (2021)
57. Hu, K., Wu, J., Schwanen, T.: Differences in energy consumption in electric vehicles: an exploratory real-world study in Beijing. J. Adv. Transp. (2017)
58. Sun, Y., Wang, S., Zhang, X., Chan, T.O., Wu, W.: Estimating local-scale domestic electricity energy consumption using demographic, nighttime light imagery and Twitter data. Energy **226**, 120351 (2021)
59. Casado-Mansilla, D., Tsolakis, A.C., Borges, C.E., Kamara-Esteban, O., Krinidis, S., Avila, J.M., et al.: Socio-economic effect on ICT-based persuasive interventions towards energy efficiency in tertiary buildings. Energies **13**(7), 1700 (2020)
60. Elakkad, N., Ismaeel, W.S.: Coupling performance-prescriptive based daylighting principles for office buildings: case study from Egypt. Ain Shams Eng. J. (2021)
61. Amasyali, K., El-Gohary, N.M.: A review of data-driven building energy consumption prediction studies. Renew. Sustain. Energy Rev. **81**, 1192–1205 (2018)
62. Bianco, V., Manca, O., Nardini, S.: Electricity consumption forecasting in Italy using linear regression models. Energy **34**(9), 1413–1421 (2009)
63. Park, D.C., El-Sharkawi, M.A., Marks, R.J., Atlas, L.E., Damborg, M.J.: Electric load forecasting using an artificial neural network. IEEE Trans. Power Syst. **6**(2), 442–449 (1991)
64. Yang, S.L., Shen, C.: A review of electric load classification in smart grid environment. Renew. Sustain. Energy Rev. **24**, 103–110 (2013)
65. Hafeez, G., Alimgeer, K.S., Wadud, Z., Shafiq, Z., Ali Khan, M.U., Khan, I., et al.: A novel accurate and fast converging deep learning-based model for electrical energy consumption forecasting in a smart grid. Energies **13**(9), 2244 (2020)
66. Syed, D., Abu-Rub, H., Ghrayeb, A., Refaat, S.S., Houchati, M., Bouhali, O., Bañales, S.: Deep learning-based short-term load forecasting approach in smart grid with clustering and consumption pattern recognition. IEEE Access **9**, 54992–55008 (2021)
67. Hafeez, G., Alimgeer, K.S., Khan, I.: Electric load forecasting based on deep learning and optimized by heuristic algorithm in smart grid. Appl. Energy **269**, 114915 (2020)
68. Hong, Y., Zhou, Y., Li, Q., Xu, W., Zheng, X.: A deep learning method for short-term residential load forecasting in smart grid. IEEE Access **8**, 55785–55797 (2020)
69. Borenstein, S.: The long-run efficiency of real-time electricity pricing. Energy J. **26**(3) (2005)
70. Oldewurtel, F., Ulbig, A., Parisio, A., Andersson, G., Morari, M.: Reducing peak electricity demand in building climate control using real-time pricing and model predictive control. In: 49th IEEE Conference on Decision and Control (CDC), pp. 1927–1932. IEEE (2010)
71. Chao, H.P.: Efficient pricing and investment in electricity markets with intermittent resources. Energy Policy **39**(7), 3945–3953 (2011)
72. Gyamfi, S., Krumdieck, S., Urmee, T.: Residential peak electricity demand response—highlights of some behavioural issues. Renew. Sustain. Energy Rev. **25**, 71–77 (2013)
73. Jiang, T., Cao, Y., Yu, L., Wang, Z.: Load shaping strategy based on energy storage and dynamic pricing in smart grid. IEEE Trans. Smart Grid **5**(6), 2868–2876 (2014)

74. Lu, R., Hong, S.H., Zhang, X.: A dynamic pricing demand response algorithm for smart grid: reinforcement learning approach. Appl. Energy **220**, 220–230 (2018)
75. Mawson, V.J., Hughes, B.R.: Deep learning techniques for energy forecasting and condition monitoring in the manufacturing sector. Energy Build. **217**, 109966 (2020)
76. Jana, R.K., Ghosh, I., Sanyal, M.K.: A granular deep learning approach for predicting energy consumption. Appl. Soft Comput. **89**, 106091 (2020)
77. Qureshi, A.S., Khan, A., Zameer, A., Usman, A.: Wind power prediction using deep neural network based meta regression and transfer learning. Appl. Soft Comput. **58**, 742–755 (2017)
78. Mujeeb, S., Alghamdi, T.A., Ullah, S., Fatima, A., Javaid, N., Saba, T.: Exploiting deep learning for wind power forecasting based on big data analytics. Appl. Sci. **9**(20), 4417 (2019)
79. Zhang, R., Feng, M., Zhang, W., Lu, S., Wang, F.: Forecast of solar energy production—a deep learning approach. In: IEEE International Conference on Big Knowledge (ICBK), pp. 73–82. IEEE (2018)
80. Almeshaiei, E., Al-Habaibeh, A., Shakmak, B.: Rapid evaluation of micro-scale photovoltaic solar energy systems using empirical methods combined with deep learning neural networks to support systems' manufacturers. J. Clean. Prod. **244**, 118788 (2020)
81. Dodiya, M., Shah, M.: A systematic study on shaping the future of solar prosumage using deep learning. Int. J. Energy Water Res. 1–11 (2021)
82. Lu, Y.S., Lai, K.Y.: Deep-learning-based power generation forecasting of thermal energy conversion. Entropy **22**(10), 1161 (2020)
83. Correa-Jullian, C., Cardemil, J.M., Droguett, E.L., Behzad, M.: Assessment of deep learning techniques for prognosis of solar thermal systems. Renewable Energy **145**, 2178–2191 (2020)
84. Lu, Y., Tian, Z., Zhou, R., Liu, W.: Multi-step-ahead prediction of thermal load in regional energy system using deep learning method. Energy Build. **233**, 110658 (2021)
85. Directive, E.C.: 98/EC of the European Parliament and of the Council, on waste and repealing certain Directives. Off. J. Eur. Union **312**, 3–30 (2008)
86. Prasanna, A., Vikash Kaushal, S.: Survey on identification and classification of waste for efficient disposal and recycling. Int. J. Eng. Technol. **7**(2.8), 520–523 (2018)
87. Saeed, M.O., Hassan, M.N., Mujeebu, M.A.: Assessment of municipal solid waste generation and recyclable materials potential in Kuala Lumpur, Malaysia. Waste manage. **29**(7), 2209–2213 (2009)
88. Yuan, H., Shen, L., Wang, J.: Major obstacles to improving the performance of waste management in China's construction industry. Facilities (2011)
89. Pattnaik, S., Reddy, M.V.: Assessment of municipal solid waste management in Puducherry (Pondicherry), India. Resour. Conserv. Recycl. **54**(8), 512–520 (2010)
90. Shinde, P.P., Oza, K.S., Kamat, R.K.: Big data predictive analysis: using R analytical tool. In: International Conference on I-SMAC (IoT in Social, Mobile, Analytics and Cloud)(I-SMAC), pp. 839–842. IEEE (2017)
91. Nowakowski, P., Pamuła, T.: Application of deep learning object classifier to improve e-waste collection planning. Waste Manage. **109**, 1–9 (2020)
92. Ahmad, S., Kim, D.H.: Quantum GIS based descriptive and predictive data analysis for effective planning of waste management. IEEE Access **8**, 46193–46205 (2020)
93. Liang, S., Gu, Y.: A deep convolutional neural network to simultaneously localize and recognize waste types in images. Waste Manage. **126**, 247–257 (2021)
94. Agarwal, H., Ahir, B., Bide, P., Jain, S., Barot, H.: Minimization of food waste in retail sector using time-series analysis and object detection algorithm. In: International Conference for Emerging Technology (INCET), pp 1–7. IEEE (2020)
95. Rashmi, G.: Regularized noise based GRU model to forecast solid waste generation in the urban region. Turk. J. Comput. Math. Educ. (TURCOMAT) **12**(10), 5449–5458 (2021)
96. Zhang, Q., Zhang, X., Mu, X., Wang, Z., Tian, R., Wang, X., Liu, X.: Recyclable waste image recognition based on deep learning. Resour. Conserv. Recycl. **171**, 105636 (2021)
97. Altikat, A., Gulbe, A., Altikat, S.: Intelligent solid waste classification using deep convolutional neural networks. Int. J. Environ. Sci. Technol. 1–8 (2021)

98. Agovino, M., Ferrara, M., Garofalo, A.: An exploratory analysis on waste management in Italy: a focus on waste disposed in landfill. Land Use Policy **57**, 669–681 (2016)
99. Knol-Kauffman, M., Solås, A.M., Arbo, P.: Government-industry dynamics in the development of offshore waste management in Norway: from prescriptive to risk-based regulation. J. Environ. Planning Manage. **64**(4), 649–670 (2021)
100. Chu, Y., Huang, C., Xie, X., Tan, B., Kamal, S., Xiong, X.: Multilayer hybrid deep-learning method for waste classification and recycling. Comput. Intell. Neurosci. **2018** (2018)
101. Sheng, T.J., Islam, M.S., Misran, N., Baharuddin, M.H., Arshad, H., Islam, M.R., et al.: An internet of things based smart waste management system using LoRa and tensorflow deep learning model. IEEE Access **8**, 148793–148811 (2020)
102. Vrancken, C., Longhurst, P., Wagland, S.: Deep learning in material recovery: development of method to create training database. Expert Syst. Appl. **125**, 268–280 (2019)
103. Cubillos, M.: Multi-site household waste generation forecasting using a deep learning approach. Waste Manage. **115**, 8–14 (2020)
104. Niska, H., Serkkola, A.: Data analytics approach to create waste generation profiles for waste management and collection. Waste Manage. **77**, 477–485 (2018)
105. Vafeiadis, T., Nizamis, A., Pavlopoulos, V., Giugliano, L., Rousopoulou, V., Ioannidis, D., Tzovaras, D.: Data analytics platform for the optimization of waste management procedures. In 15th International Conference on Distributed Computing in Sensor Systems (DCOSS), pp. 333–338. IEEE (2019)
106. Xia, M., Li, T., Xu, L., Liu, L., De Silva, C.W.: Fault diagnosis for rotating machinery using multiple sensors and convolutional neural networks. IEEE/ASME Trans. Mechatron. **23**(1), 101–110 (2017)
107. Massaoudi, M., Abu-Rub, H., Refaat, S.S., Chihi, I., Oueslati, F.S.: Deep learning in smart grid technology: a review of recent advancements and future prospects. IEEE Access **9**, 54558–54578 (2021)

Chapter 3
QoS Aware Service Provisioning and Resource Distribution in 4G/5G Heterogeneous Networks

Rintu Nath

Introduction

The 4G technology gives enhanced features to mobile communication and foster creation of associated applications and service ecosystem to revolutionize ICT. Growth of Internet of Things (IoT) is generating high volume data and causing congestion of network. Unpredictable data flow from billions of connected devices will affect throughput and may cause deterioration of the Quality of Service (QoS) [1]. Due to large scale deployment of IoT technologies and ever-increasing number of mobile users on 4G and 5G networks, increasing channel efficiency while ensuring QoS constraints is challenging. The new age mobile communication system is driven by heterogeneous applications, variable user requirement and unpredictable data rate. Deployment of services with guaranteed QoS and increased spectrum efficiency are often conflicting and, hence challenging.

Advancement in 5G network is expected to increase spectrum efficiency with higher throughput. However, design constraints for QoS provisioning need to be addressed for a reliable 5G network. Architecture, framework, and scheduling algorithm imposing new design challenges for 5G. Delay bounded QoS requirements for high volume data is difficult to fulfill in a 4G network. To characterize QoS for delay bound high-volume data, homogeneous statistical provisioning is done that guarantees QoS for each link. However, for a 5G network, different delay bound heterogeneous data traffic having different QoS constraints can be guaranteed.

R. Nath (✉)
Vigyan Prasar, Department of Science and Technology, AI Building, Technology Bhawan, New Mehrauli Road, New Delhi 110016, India
e-mail: Rnath@vigyanprasar.gov.in

When a network is configured for a large buffer, it becomes unstable and subsequent delay in packet-switched network cause bufferbloat. Variation in packet delay causes jitter and overall throughput reduces. Applications like online gaming, Voice over IP (VoIP), online transactions become unreliable due to bufferbloat and jitter. To overcome these problems and to ensure QoS, several solutions exist. Solutions are primarily categorized in two groups, end-to-end and in-Network solutions. Congestion Control Algorithms (CCA) in 4G and 5G networks try to ensure high throughput and low latency. However, CCAs need to have fairness while interacting with different networks and should only be deployed after QoS parameters are ensured.

The aim of this chapter is to discuss various service provisioning techniques and throughput constraint resource allocation that guarantees QoS for a diverse set of services, applications, and user requirement in 4G and 5G networks. Simulation of heterogeneous statistical delay bounded QoS provisioning with differential baud rate and fading parameters are done and results discussed.

Related Work

Challenges of 4G and 5G networking for QoS provisioning, resource allocation, information flow, coding and modulation schemes, and resource management are listed in [2, 3].

There is a fundamental tradeoff between QoS provisioning and effective throughput. Tang et al. [4] reported power and rate adaptation scheme to maximize system throughput for a given delay QoS constraint. The scheme is applied on a block fading channel model in which higher channel correlation gives faster convergence of power-control policy and stringent adaptation of QoS. Call Admission Control (CAC)-based QoS provisioning in a resource management framework is reported by Inaba et al. [5]. Beshley et al. [6] proposed another resource management framework with end-to-end QoS management algorithm in 4G/5G networks. The authors proposed a modified architecture of Long-Term Evolution (LTE) for IoT services. Haile et al. [7] discussed an end-to-end congestion control approach for 4G and 5G networks that guarantees high throughput. Various dynamic network slicing and resource allocation techniques are presented. The authors, however, did not present any data on overall spectrum efficiency with QoS provisioning.

Beshay et al. [8] presented a framework for link-coupled TCP for 5G networks. It is a transport layer solution that can take advantage of 5G architecture trends. Without modifying TCP clients, LCTCP can be deployed. However, network latency and associated bottleneck is not addressed by LCTCP. A machine learning-based 5G architecture is proposed by Zhu et al. [9]. The authors proposed supervised learning-based QoS assurance protocols to mitigate network congestions and to improve channel throughput. Park et al. [10] proposed an improved version of congestion control algorithm ExLL The primary objective of the algorithm is to ensure congestion reduction and considers latency as one of the QoS constraints. For downlink packet reception, ExLL utilizes cellular bandwidth inference, and need minimum round-trip time (RTT). One of the drawbacks of ExLL is that it is not able to improve channel throughput effectively. Improvement in channel throughput by statistical-QoS driven resource allocation is reported by Zhang et al. [11]. The proposed allocation policies are applicable in both asymptotic and non-asymptotic regimes of mmWave-based 5G network.

Xie et al. [12] proposed PBE-CC algorithm for congestion control that improves channel throughput significantly by precise measurement of rise and fall of wireless capacity demand. Another end-to-end congestion control approach is reported by Haile et al. [13]. The authors presented a congestion control algorithm that ensures low latency and high throughput in a highly variable network links. In addition, the authors discussed deployability of the algorithm. A service oriented transmission protocol in 5G network is discussed by Chen et al. [14]. Sharma et al. [15] presented a review of wireless backhaul networks and emerging trends of 5G networks. Ahmad et al. [16] discussed QoS constrained dynamic spectrum sharing using cognitive radio networks.

QoS Provisioning for M2M

The 4th Generation Long-Term Evolution (LTE) wireless network has dominant share of uplink data for Machine to Machine (M2M) communication. LTE uses shared radio channel based Single-Carrier Frequency Division Multiple Access (SC-FDMA) technique for uplink and Orthogonal FDMA (OFDMA) for downlink for M2M terminals. The focus of this section is on time domain scheduling using QoS class Identifier (QCI).

Enhanced LTE Network Architecture for 5G Networks

The existing LTE architecture of 4G network is not able to handle growing data rate of M2M communication. This chapter gives an overview of the improved architecture for 5G network that enable mass deployment of M2M and H2M data.

5G network is expected to cater three types of user classes, namely, massive Machine Type Communications (mMTC), enhanced Mobile Broad Band (eMBB), and Ultra-Reliable and Low Latency Communications (URLLC). Applications of mMTC include smart cities, smart homes, and office automation. Online gaming, ultra-high definition videos need eMBB support of 5G. Applications like autonomous vehicles, robots need reliable communication network and 5G URLLC is expected to deliver that.

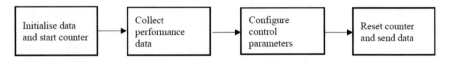

Fig. 3.1 Generalized congestion control algorithm flowchart

End-to-end cellular congestion control algorithms may be classified in three cate-gories (i) predictive algorithm, (ii) reactive algorithm, and (iii) network assisted algorithm. Predictive algorithms may be fixed or dynamic interval. Reactive algo-rithms may be based on loss, delay, or rate triggered. Network assisted algorithms are based on in-band or out-band signaling. A generalized congestion control algorithm flowchart is illustrated in Fig. 3.1. For CCA, data is initialized, and counter starts. In the next stage, performance data is collected, and control parameters are config-ured. Finally, counter is reset, and data sent. QoS constrained congestion control algorithms are mentioned in Table 3.1.

Two primary constraints of CCAs are cross traffic and cellular bottleneck. Cross traffic is reduced by implementing per-user-queue. Cellular bottleneck is reduced by moving content nearer to the end users. Moving content closer to cellular access link reduces cost of bandwidth and reduces delay.

Resource Allocation to Ensure QoS

Link adaptation, admission control, removal and handover management, packet switching are some of the key features of resource management. For LTE and LTE-A, packet switching is done at MAC layer with the objective of increasing spectral efficiency with larger throughput. Scheduling decision is based on QoS requirement and Channel State Information (CSI).

Mobile devices should be able to connect to multiple access points, e.g., 5G, WLAN, LTE. Dual connectivity can exploit diversities of access points and deliver high data rate. Another important concept of resource allocation is network slicing by dividing physical network in multiple logical entities. Network slicing enables dynamic allocation of resources with different use cases.

Adaptive Channel Bandwidth Selection in LTE 4G/5G Networks

For fading channels 3G communication, generalized finite-state Markov channel (FSMC) is useful. However, for 5G network, Filter Bank Multi Carrier (FBMC) is utilized. Orthogonal Frequency Division Multiplexing (OFDM) is primarily based on FBMC.

Table 3.1 QoS constrained congestion control algorithms (CCAs)

Algorithm	Type	Features	Design criteria	Applications
C2TCP [17]	End-to-end solution	Delay sensitive, network state profiling not required	Sits on top of loss-based TCP	High throughput and low delay applications
X-TCP [18]	5G networks and 3GPP new radio	Provides large bandwidth, increased cell throughput	High variability	Congestion control with TCP CUBIC
PBE-CC [12]	Physical-layer bandwidth measurements	5G new radio innovations, increased wireless capacity	Cellular aware, congestion control protocols	Software-defined radios, PBE sender and receiver
DL-TCP [19]	Deep-learning-based TCP	TCP congestion window adjustment	Mobility information, signal strength	5G mmWave network
LCTCP [8]	Link-coupled TCP	Transport layer solution	Isolation of 5G access link, lightweight signaling	Link buffer, application server
DRWA [20]	Dynamic receiver window adjustment	Solves bufferbloat problem, reduces latency	TCP modification	Over the air (OTA) updates
CQIC [21]	Cross-layer congestion control	Physical layer information exchange	Adjustment in packet sending behavior	Cross-layer optimization
QTCP [22]	Reinforcement Q-learning framework	High throughput, low transmission latency	Hard-coded rules not required	Optimal congestion control
CDBE [23]	Client driven bandwidth estimation	Down-stream performance enhancement	Varying cellular BW	Variation of down-stream delay
ExLL [10]	Extremely low latency congestion control	Controls congestion window	Continuous bandwidth probing	Dynamic cellular channels

System Architecture

To overcome bufferbloat and subsequent jitter, a point-to-point Congestion Control Algorithms (CCA)-based solution is preferred. The system architecture is illustrated in Fig. 3.2. Distributed data and per-user-data are the two main reasons of bottleneck. CCA feedback ensures lesser congestion in network traffic. Data source consists of datalink layer. Theses datalink layers are stored in input buffer. In the next stage, physical layer splits datalink layers into bit streams. Based on QoS constraints and

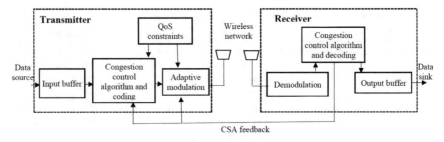

Fig. 3.2 System architecture

CCA feedback, adaptive modulation is done. In the receiving end reverse process is followed to reconstruct original message signal. Demodulation consists of an intermediate frequency stage that reduces modulated signal frequency.

TCP/IP transport layer supports CCAs. This signify that transport layer information and QoS constraints are the deciding factors for bottleneck link access. This in turn has preferential selection for top layer applications. Channel State Information (CSI) is updated based on the feedback received from receiver. Deployment of this model is possible even for applications where link layer information is not accessible. CSI of some applications may be dependent of APIs to interact with lower levels. API interaction allows continuous information flow for CCAs. However, ties of CSI and API may be loosely bound or tightly bound, depending on specific technologies for which applications are designed. Coordination among different functional blocks of transmitter and receiver is important for effective communication channel that is devoid of any bottleneck. CSA feedback introduce modifications in the input stack and subsequently all the down-stream layers of transmitter. Similarly, any modification at the receiving end would update network stack of users.

Delay Bounded QoS Provisioning

For delay bound Content Delivery Networks (CDN), performance tradeoff is done between available capacity and bit rate. Larger queue is required when capacity is increased. Underutilization of capacity reduces throughput. In some cases, some minimum delay may have to be guaranteed, while CSI may not be able to provide that delay. Hence CCAs prioritizes QoS delay constraint through underutilization of queue stack. CSI mostly receives information like delay, loss and packet count from transport layer. Based on CSI feedback, congestion control algorithms changes control decisions. However, delay due to adaptive provisioning of lower layers may cause bottleneck. In such cases, transport layer will not be able to differentiate between delay caused by link rescheduling and queuing delay. In such cases cross-layer feedback, i.e., feedback received from network stack becomes dependable and gaining popularity in 5G networks. This approach also gives a reliable estimate about congestion.

Hybrid Scheduler with QoS Class Identifier

M2M communication model using LTE wireless network communicate with several servers. Hence data flow is more in the uplink. In general, M2M traffic is delay tolerant and designed for end-to-end QoS. Till 3G network, there was no specific QoS class for M2M communication. However, 4G and 5G network introduced QoS Class Identifier (QCI), which is used for admission control, queuing decision and congestion control algorithm. Shared radio channels between LTE access network and user equipment communicate by OFDMA and SC-FDMA. Admission control, mobility monitoring, and resource allocation are done via enhanced Node B (eNodeB). Both time domain and frequency domain subcarrier allocation are done for LTE subscribers. Optimal data packet set for transmission is time domain scheduling, whereas mapping data packets to resource blocks is frequency domain scheduling.

Conclusion

A delay bound QoS provisioning technique and performance tradeoff for content delivery networks is discussed in the chapter. Large input buffer causes bufferbloat and subsequent jittering the network. Congestion control algorithm along with Channel State Information (CSI) feedback plays important role to mitigate bottle-neck and guarantee QoS. Further research on efficient congestion control algorithm, delay bounded QoS provisioning, effective protocols for steering antenna beam-forming, and improved error correction coding will make 5G network more useful and user friendly. Advancement in technologies like, Cloud Radio Access Networks (C-RAN), Heterogeneous C-RAN, Software-Defined Networking (SDN), Massive MIMO, Self-Organizing Network (SON), and opportunistic network will lead to a new generation of communication network. Energy efficiency should ideally be one of the important constraints toward a sustainable 5G network. It is almost impossible for a single technology to meet all criteria and converge—hence, a sequential roll-out of 5G will be able to meet-up with the user expectations and will be sustainable. Cooperation and coordination among industries, academia and regulatory bodies would play an important role in balancing QoS, energy efficiency and effective spectrum utilization.

References

1. Sharma, T., Chehri, A., Fortier, P.: Review of optical and wireless backhaul networks and emerging trends of next generation 5G and 6G technologies. Trans. Emerg. Telecommun. Technol. **32**(3), 1–16 (2021). https://doi.org/10.1002/ett.4155
2. Abdalla, I., Venkatesan, S.: A QoE preserving M2M-aware hybrid scheduler for LTE uplink. In: International Conference on Selected Topics in Mobile and Wireless Networking (MoWNeT), vol. 7, pp. 127–132 (2013). https://doi.org/10.1109/MoWNet.2013.6613808
3. Akhtar, T., Tselios, C., Politis, I.: Radio resource management: approaches and implementations from 4G to 5G and beyond. **27**(1) (2021)
4. Tang, J., Zhang, X.: Quality-of-service driven power and rate adaptation for multichannel communications over wireless links. IEEE Trans. Wirel. Commun. **6**(12), 4349–4360 (2007). https://doi.org/10.1109/TWC.2007.06031
5. Inaba, T., Sakamoto, S., Oda, T., Barolli, L., Takizawa, M.: A new FACS for cellular wireless networks considering QoS: a comparison study of FuzzyC with MATLAB. In: Proceedings of the 18th International Conference on Network-Based Information Systems. NBiS 2015, pp. 338–344 (2015). https://doi.org/10.1109/NBiS.2015.52
6. Beshley, M., Kryvinska, N., Seliuchenko, M., Beshley, H., Shakshuki, E.M., Yasar, A.U.H.: End-to-end QoS 'Smart Queue' management algorithms and traffic prioritization mechanisms for narrow-band internet of things services in 4G/5G networks. Sensors (Switzerland) **20**(8) (2020). https://doi.org/10.3390/s20082324
7. Haile, H., Grinnemo, K.J., Ferlin, S., Hurtig, P., Brunstrom, A.: End-to-end congestion control approaches for high throughput and low delay in 4G/5G cellular networks. Comput. Networks **186**, 107692 (2021). https://doi.org/10.1016/j.comnet.2020.107692
8. Beshay, J.D., Nasrabadi, A.T., Prakash, R., Francini, A.: Link-coupled TCP for 5G networks. In: IEEE/ACM 25th International Symposium on Quality of Service (IWQoS) (2017). https://doi.org/10.1109/IWQoS.2017.7969170
9. Zhu, G., Zan, J., Yang, Y., Qi, X.: A supervised learning based QoS assurance architecture for 5G networks. IEEE Access **7**, 43598–43606 (2019). https://doi.org/10.1109/ACCESS.2019.2907142
10. Park, S., Lee, J., Kim, J., Lee, J., Lee, K.: ExLL: an extremely low-latency congestion control for mobile cellular networks. In: CoNEXT 2020: Proceedings of the 16th International Conference on emerging Networking EXperiments and Technologies, pp. 307–319 (2020)
11. Zhang, X., Wang, J., Poor, H.V.: Heterogeneous statistical-QoS driven resource allocation over mmWave massive-MIMO based 5G mobile wireless networks in the non-asymptotic regime. IEEE J. Sel. Areas Commun. **37**(12), 2727–2743 (2019). https://doi.org/10.1109/JSAC.2019.2947941
12. Xie, Y., Yi, F., Jamieson, K.: PBE-CC: congestion control via endpoint-centric, physical-layer bandwidth measurements. In: SIGCOMM 2020: Proceedings of the Annual Conference of the ACM Special Interest Group on Data Communication on the Applications, Technologies, Architectures, and Protocols for Computer Communication, pp. 451–464 (2020). https://doi.org/10.1145/3387514.3405880
13. Haile, H., Grinnemo, K.J., Ferlin, S., Hurtig, P., Brunstrom, A.: End-to-end congestion control approaches for high throughput and low delay in 4G/5G cellular networks. Comput. Networks **186**, 107692 (2021). https://doi.org/10.1016/j.comnet.2020.107692
14. Chen, J., et al.: SDATP: an SDN-based traffic-adaptive and service-oriented transmission protocol. IEEE Trans. Cogn. Commun. Netw. **6**(2), 756–770 (2020). https://doi.org/10.1109/TCCN.2019.2963149
15. Sharma, T., Chehri, A., Fortier, P.: Review of optical and wireless backhaul networks and emerging trends of next generation 5G and 6G technologies. Trans. Emerg. Telecommun. Technol. **32**(3), 1–16 (2021). https://doi.org/10.1002/ett.4155
16. Ahmad, W.S.H.M.W., et al.: 5G technology: towards dynamic spectrum sharing using cognitive radio networks. IEEE Access **8**, 14460–14488 (2020). https://doi.org/10.1109/ACCESS.2020.2966271

17. Abbasloo, S., Li, T., Xu, Y., Chao, H.J.: Cellular controlled delay TCP (C2TCP). In: IFIP Networking Conference (IFIP Networking) and Workshops, pp 118–126 (2018). https://doi. org/10.23919/IFIPNetworking.2018.8696844
18. Azzino, T., Drago, M., Polese, M., Zanella, A., Zorzi, M.: X-TCP: a cross layer approach for TCP uplink flows in mmwave networks. In: 16th Annual Mediterranean Ad Hoc Networking Workshop (Med-Hoc-Net), pp. 1–6 (2017). https://doi.org/10.1109/MedHocNet.2017.800 1650
19. Na, W., Bae, B., Cho, S., Kim, N.: DL-TCP: deep learning-based transmission control protocol for disaster 5G mmWave networks. IEEE Access 7, 145134–145144 (2019). https://doi.org/ 10.1109/ACCESS.2019.2945582
20. Jiang, H., Wang, Y., Lee, K., Rhee, I.: DRWA: a receiver-centric solution to bufferbloat in cellular networks. IEEE Trans. Mob. Comput. 15(11), 2719–2734 (2016). https://doi.org/10. 1109/TMC.2015.2510641
21. Lu, F., Du, H., Jain, A., Voelker, G.M., Snoeren, A.C., Terzis, A.: CQIC: revisiting cross-layer congestion control for cellular networks. In: HotMobile 2015: Proceedings of the 16th International Workshop on Mobile Computing Systems and Applications, pp. 45–50 (2015). https://doi.org/10.1145/2699343.2699345
22. Li, W., Zhou, F., Chowdhury, K.R., Meleis, W.M.: QTCP: adaptive congestion control with reinforcement learning. IEEE Trans. Netw. Sci. Eng. 4697, 1–13 (2018). https://doi.org/10. 1109/TNSE.2018.2835758
23. Zhong, Z., Hamchaoui, I., Ferrieux, A., Khatoun, R., Serhrouchni, A.: CDBE: a cooperative way to improve end-to-end congestion control in mobile network. In: International Conference on Wireless and Mobile Computing, Networking and Communications (WiMob), vol. 2018, pp. 216–223 (2018). https://doi.org/10.1109/WiMOB.2018.8589175

Chapter 4
Leveraging Fog Computing for Healthcare

Avita Katal

Introduction

Consumers and organizations are becoming more dependent on PCs and digital phones to do daily activities. Information is produced by such devices via a variety of sensors and functions. As a result, businesses produce and store vast amounts of data on a regular basis. The amount of data collected by sensors has risen dramatically since the introduction of IoT. Big data analysis has received a lot of attention in past times because of enormous increase in the amount of data generated and the difficulties of conventional systems in handling various forms of unstructured and structured data [1]. Every business is increasingly emphasizing data analysis to obtain relevant insights in order to make critical choices. Organizations today demand a flexible IT architecture as a result of cloud computing's connectivity, flexibility, and pay-per-use features. However, big data, or data generated by billions of sensors, cannot be transmitted and analyzed on the cloud. Furthermore, certain IoT applications must be handled quicker than the cloud's existing capacity. The fog computing (FC) concept, which uses the computational capability of equipment located near clients (idle computing power) to enable storage, compute, and communication at the edge, can assist in addressing this issue.

FC is a distributed computing model that acts as a link among data centers and the Internet of Things (IoT). It enables cloud-based solutions to be moved closer to IoT devices/sensors by providing computing, networking, and data storage [2]. In 2012, Cisco introduced the concept of FC to address the constraints of IoT applications in traditional environments. IoT devices/sensors, and also real and latency-sensitive available services, are widely spread near the network's edge. Since cloud data centers

A. Katal (✉)
School of Computer Science, University of Petroleum and Energy Studies, Dehradun,
Uttarakhand, India
e-mail: avita207@gmail.com

© The Author(s), under exclusive license to Springer Nature Singapore Pte Ltd. 2023   51
V. Kadyan et al. (eds.), *Deep Learning Technologies for the Sustainable
Development Goals*, Advanced Technologies and Societal Change,
https://doi.org/10.1007/978-981-19-5723-9_4

are widely dispersed, they typically fail to fulfill the memory and networking requirements of billions of globally scattered IoT devices. As a consequence, the infrastructure is overburdened, there is substantial delay in delivery of services, and the quality of service (QoS) is poor.

A FC environment is frequently composed of ordinary network equipment such as switches, routers, cable boxes, web servers, base stations (BS), and so on, and it can be located in areas near to IoT devices. These components are equipped with a range of computing, memory, communication, and other characteristics as well as the capability to simplify the performance of operations. FC, as a virtue of its networking devices, is capable of generating a wide regional spread of cloud-based operations. In addition, FC assists in position recognition, mobility support, real-time engagements, flexibility, and compatibility. In terms of operation delay, energy usage, traffic patterns, costs and capital expenditures, FC can be inexpensive. In this regard, FC outperforms cloud computing (CC) in terms of meeting the requirements for IoT applications.

As Topol argues in "The Creative Destruction of Medicine" [3], healthcare is about to undergo its most profound transformations in its history. Healthcare 4.0 is a follow-up to earlier healthcare versions that was inspired by Industry 4.0. Its goal is to unite a patient-centered system by developing an enhanced virtualized and personalized healthcare domain. It is based on the exchange of patient records between healthcare providers via electronic health record (EHR) repositories. Data sharing has improved clinicians' access to data from any place. It also makes it easier for doctors to exchange their patients' data with their colleagues in order to make the most accurate diagnosis and arrange the best possible therapy. Healthcare 4.0 has successfully connected consumers and health systems through technology. Nevertheless, data exchange has brought additional concerns such as data protection, data governance, networking devices, data verification, and so on. Aside from this, updated technology needs skill development in order to be used effectively. Wireless sensor technology is one of the forces driving these developments. Sensors are becoming smaller, allowing them to be worn without interfering with daily life while still providing access to an expanding number of biometric characteristics [4]. This is critical when data must be collected on a continual basis. The BioStamp [5], for example, is a band-aid-sized sensor that can monitor numerous biometric signals and is easily applied to the skin. Kang et al. [6] go on to describe an improved process for imprinting sensors precisely onto skin-applying sticky film. A variety of biometrics may also be detected using contact lenses. Such improvements pave the way for people to be equipped with hundreds of devices. There is also a plethora of fitness trackers available. They foreshadow a future where every individual, irrespective of condition, is continually monitored. Nevertheless, such a straightforward sensor-to-cloud design is impractical for many uses in healthcare management. In some cases, regulations prevent the keeping of patient information outside the clinic. Dependence on faraway data centers is particularly unacceptable for various uses due to patient care in the case of network and data center outages. In health informatics, fog computing is one method for overcoming the gap among sensors and analytics.

This is a type of decentralized network structure in which application-specific information is stored not just in data centers (the cloud) or gadgets close to consumers, but also in the infrastructure elements that connect them.

Architecture of Fog Computing

The standard design of FC architecture is an essential researched topic. Several FC concepts have been proposed in recent times. They are primarily based on the fundamental three-layer architecture. FC brings cloud storage service to edge devices by building a fog layer between the end source and the network. Figure 4.1 shows the fog architecture. The three levels of the hierarchical architecture are as follows:

- Terminal layer: The terminal layer is the tier that is closest to the end consumer and the actual surroundings. It is composed of many IoT devices such as sensing devices, cell phones, smart vehicles and readers, among others. Despite the fact that cell phones and smart vehicles have computing capabilities, users simply use these as smart sensing devices in this situation. In practice, these devices are widely scattered regionally. They are capable of collecting actual thing or event characteristic information and transmitting it to the upper layer for storage and processing.

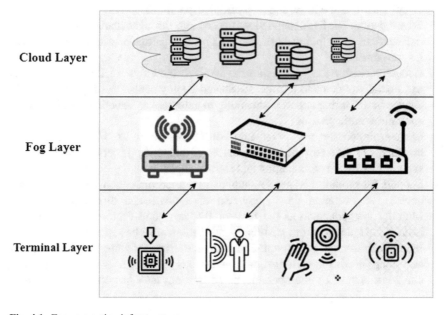

Fig. 4.1 Fog computing infrastructure

- Fog layer: This tier is located at the network's perimeter. The FC layer comprises of several fog nodes such as gateways, routers, converters, access points, base stations, specialized fog computers, and so on. Such fog nodes are widely spread among end sources and the network, including cafes, retail malls, bus stations, streets, and parks. They could be fixed in one location or mobile on a mobile carriage. To obtain functions, end devices may easily link to fog nodes. They can analyze, send, and store large observed data. Actual monitoring and delay sensitive operations are feasible in the fog layer. Additionally, the fog nodes are connected to the data center via an IP network infrastructure and are responsible for communicating and collaborating with cloud in order to obtain more robust processing facilities.
- Cloud layer: The CC level consists of several elevated storage devices that provide a wide range of software solutions such as home automation, smart cities, smart factory, and so on. It has considerable computational and memory abilities, allowing for detailed computation evaluation and lengthy data management.

Characteristics of Fog Computing

- Heterogeneity: FC is a virtualized framework that delivers storage, networking, and processing operations among end nodes and standard CC data centers, which are often not located at the network's edge. The cloud and the fog are both based on computation, memory, and networking capabilities.
- Edge location: The fog's foundations may be seen throughout ambitions to provide rich services to endpoints at the network's edge, particularly applications needs less response time.
- Geographical distribution: Despite the relatively cloud system, the fog's functions and operations demand a huge distributed deployments. The fog will actively engage in delivering superior streaming to automobiles traveling over highways and railroads via proxies [7].
- Large-scale sensor networks: Large-scale sensor nodes for environmental tracking, and the smart grid, are instances of inherently dispersed processes that support distributed computation and storage capabilities.
- Support for mobility: Many fog applications need direct contact with portable devices, necessitating the deployment of a distributed directories network. Mobility methods such as the Locator ID Separation Protocol (LISP), which isolates host identity from position identification, must be supported.
- Interoperability and federation: The smooth delivery of some services (such as streaming) needs the participation of numerous vendors. As a consequence, fog components must be interoperable, and operations must be pooled across zones.

Applications of Fog Computing

FC is highly adapted to low-latency applications [8], as a result, it may be utilized in any delay sensitive operation, such as medical, crisis, and malware systems. Here are a few examples of FC applications:

- Smart grid: It is the future network for the distribution of electricity. Smart grids include power stations, generators, converters, and other equipment. It builds an automatic and decentralized energy transmission infrastructure using simultaneous energy and information flows. The smart grid is a visible power generation system that allows clients and industry suppliers to keep a check on billing, output, and real-time usage. In the systems of big data, billions of smart meters are put in consumer homes. Fog detectors are used at the edge of the system to collect, evaluate, and filter data before transferring it to a cloud data center for long-term retention.

- Healthcare system: Medical software systems are sluggish to react and provide critical health information. The produced data contains crucial data. Likewise, location information may be problematic in some cases. In telehealth and telemedicine applications, increased instability and latency can generate a variety of issues. In such cases, fog computing may be appropriate for the healthcare ecosystem. Due to the low latency connected with embedded health devices, ambulance connectivity, and transportable connectivity to client medical data, FC plays an important role in essential healthcare operations [9]. In [10], the author proposed a technique for brain patients. FC is used in the proposed system to identify, forecast, and prevent stroke victims from collapsing. They applied an object tracking training approach to both edge devices and cloud services.

- Augmented reality: The capacity to surround and overlay digital and virtual things into the actual environment is referred to as augmented reality. Augmented reality data has low delay and a higher data processing rate to offer accurate data depending on the client's position [11]. Delays are highly important in mixed reality systems. Small delays in responding may jeopardize the client's abilities. AR techniques require use of image processing algorithms to assess actual frames of videos while still evaluating other inputs such as speech, and eventually to create relevant appropriate material on the device. In any event, humans are very alert to lags in a series of subsequent contacts. Even a few milliseconds of processing latency impairs the user engagement and leads to a negative client response. As a consequence, augmented reality devices benefit from FC by boosting throughput while decreasing computational latencies [12].

- Traffic control system: A video camera in vehicle communication that recognizes the bright colors of an ambulance can quickly change the lighting system and clear the track for the car to pass the congestion. Smart street lights use sensors to identify the existence of people and bicycles, and the speed and distance of approaching automobiles. Aside from that smart lighting switches on autonomously when the sensor detects motion and switches off as the vehicle passes. Adjacent smart lights that act as fog nodes collaborate to make a traffic signal green and alert approaching automobiles. A traffic management system may help avoid accidents, maintain steady traffic, and gather important data for the evaluation and development of a product's efficiency [13].

Fog Computing in Healthcare

Most nations' healthcare systems confront significant problems, which will worsen as the population ages and chronic diseases become more prevalent. Many countries are also facing a nursing personnel shortage. Likewise, there is a desire to keep expenses down while still offering people with high care. As a consequence, the medical field promotes a data-driven approach to healthcare. Individuals can be monitored remotely as component of this delivery model that improves mobility, safety, economy, and continuation of diagnosis for people while also reducing total healthcare costs. Today, hospitals lose a lot of time manually measuring biometric indicators and transmitting data across systems, which typically involves pen and paper. Caretakers will have more time because of remote monitoring. Other enhancements include automatic supervision, which may be used in place of manual monitoring. Bertini et al. [14] explain the benefits of monitoring system over in hospital follow-ups, particularly a possible increase in survival. One point of emphasis is the improvement of hospital procedures. There are various processes to follow. Many stages are planned manually and so completed in a chronological manner instead of making better use of facilities. Topol [3] refers to this as "digitizing humanity." One requirement is that data not be handled in silos but instead be combined with other resources and seen in perspective. One trend is a radical departure from reactive therapy, which involves admitting people to a clinic only after an incident happens, and toward more preventive care [15]. This starts with screening normal individuals to keep them out of the hospital as long as possible. Improving patients' capacity to be monitored at home helps them to be discharged from the hospital sooner. In principle, this means that the distinctions between hospitals, homes, and other sites of care are getting increasingly unclear: Medicine is available at all times and in all places.

Current advancements in ICTs, such as the IoT and cyber-physical systems (CPS), allow us to build more smart and predictive medical services, in both everyday life (home/office) and in clinics. The bulk of IoT-based health systems, notably smart homes or clinics, need a bridge point (i.e., gateway) among the sensing equipment devices and the Internet. The authors in [16] strategically position gateways at the network's edge to provide a range of higher-level capabilities such as local storage, real-time local data analysis, integrated data mining, and so on, resulting in a smart eHealth gateway. The researchers then recommended applying the concept of fog computing to medical IoT network by building a spatial intermediary layer of intelligence among sensor network and the cloud. By assuming on some of the tasks of the sensor nodes and a remote healthcare center, their fog-assisted developed system addressed numerous issues in ubiquitous health systems, including mobility, power efficiency, adaptability, and reliability. Furthermore, the authors in [17] provides an efficient centralized safe framework for end-to-end connectivity of a cloud-hosted IoT-based health service The software is run utilizing the fog computing environment on the suggested architecture. The medical information is collected through sensors and securely transmitted to near-edge gadgets. Finally, the data is uploaded to the cloud for simple access by healthcare professionals. The suggested system employs asynchronous communication between cloud-deployed apps and data servers.

Numerous sensors and equipment generate enormous volumes of data in healthcare applications, which are the focus of essential activities. Their management at the network's edge can be accomplished with fog computing. The authors of Mutlag et al. [18] proposed a multi-agent fog computing paradigm for managing healthcare essential activities. The multi-agent system's primary function is to map between three decision tables in order to maximize scheduling essential activities by allocating tasks based on their priority, network load, and network resource availability.

The fundamental CC itself is not the best solution for changing healthcare challenges, such as patient care delays. Moreover, the medical care and apps that are directly available on CC do not satisfy the criteria of the Healthcare 4.0 environment. It has certain disadvantages, such as sluggish real-time reaction and latency. A minor hiccup in healthcare might cost a patient his or her life: hence, FC has entered the scene to improve services and applications. FC provides on-time and consistent delivery of services despite addressing difficulties like delays or jitters, as well as cost overheads associated with data transmission to the cloud. It is a distributed flat design that increases storage, processing, and development platforms throughout the CC.

Rapid advancements in biological sensing, less power and low-cost circuits, and wireless networks have brought this notion closer to reality. However, significant issues and difficulties like storing the data, security and privacy of the medical data must still be addressed. When it comes to continuous human health observation, tracking systems will be the first to capture all signals from the body.

Icker Institute in 1988 design the patient centric care (PCC) [19] is one of the emerging healthcare approaches. This strategy was centered on patients and meeting their healthcare requirements. Healthcare 4.0 includes all already present medical centers, including major hospital companies such as institutions, (ii) minor health professionals such as clinicians and pharmacies, and (iii) health centers and pharmacies, and (iii) non-clinical settings, such as outpatient care facilities, care homes, and rural areas with limited medical care. Telehealth refers to potential sectors in which FC might play a significant role in providing care to institutions, pharmacies, and remote places [20].

Healthcare Application Needs

CC can provide pervasive, on-demand, and accessible network access to digital capabilities that can be distributed and provided immediately with no human effort or network operator management. These devices include properties of both clusters and grids, and also distinctive qualities and abilities, such as robust virtualization. As a result, such frameworks make it easier to provide third-party, value-added solutions by utilizing computing, memory, and system software without generating the necessary hosting infrastructure. Many health detection systems have relied on remote cloud services to collect and manage massive amounts of data generated by IoT devices. There are, however, certain difficulties with connectivity delay, position identification, and massive data transmission. As healthcare systems get wider and increasingly sophisticated, the danger of storage and communication mistakes increases, with a minor mistake in the collected data leading to incorrect treatment recommendations and potentially putting a human's life in jeopardy. In medical settings, FC is one method of overcoming the barrier among sensors and statistics.

FC is a scalable architecture that places storing and processing near-edge devices. Users are compensated for making a part of their equipment available to host these services. FC, as defined by Iorga et al. [21], is a multilayered infrastructure that provides pervasive connectivity to a common continuity of flexible computational power. The idea, which is built on physical and virtual fog nodes placed among connected phones and central operations, simplifies the implementation of latency-aware networked infrastructure and technologies.

- *Bandwidth*—The amount of connections, the analog-to-digital converter (ADC) normalization step-size in bits, and the sample rate all influence the information rates of distinct medical information [22]. Internal temperature, for example, requires just a sample rate of 0.2 Hz. This leads to a bandwidth of 2.4 bit/s when just a 12-bit ADC is used [23]. A 12-bit ADC was used to detect blood pressure at 120 Hz at 1.44 kbit/s [23]. For pulse oximetry testing, 600 Hz sampling and 7.2 kbit/s are required [23]. ECGs (electrocardiograms) usually necessitate the use of more than one lead. For medical uses, a 5-lead ECG needs 36–216 kbit/s, based on the pulse width and track record [22, 24]. Electromyograms (EMG) are electrical impulses produced by muscle that can be used for a range of purposes, such as food chewing identification. Minimum bandwidth requirements for such usage scenarios are 20.48 kbit/s and 96 kbit/s, respectively. The electroencephalogram (EEG) uses a large number of leads to detect electrical activity in the brain. An EEG with 192 leads can take up to 921.6 kbit/s of bandwidth. This illustrates how much variation there is in the bitrates of biomedical parameters.
- *Latency*—The needs for delay differ considerably based on the planned use for the information. In experiments with cardiologists, the authors in [24] showed that delays of up to 2–4 s in real-time surveillance are tolerable for ECG. From a technological standpoint, they are rather lax criteria. Stricter criteria are required for Tactile Internet applications, such as the control of exoskeletons that allow paralyzed people to walk. Other examples of latency limitations come from programs of telehealth that operate in remote areas with insufficient networking connectivity.
- *Energy efficiency*—Since replacing cells hinders sensor use, energy conservation is a major issue. Although some in-body monitors gather energy via heat or motion activity, other gadgets may necessitate client surgery to refresh a battery.

- *Dependability*—System failures have varying effects depending on what data is utilized for, ranging from small annoyance to a significant danger to the lives of the patients. As a result, consistency is probably the most important factor to examine, and it is intimately related to security threat resistance.
- *Security*—The security standards in healthcare are stringent because of the complexity of data of the patients and the possible disastrous consequences of interfered with or altered devices and systems. In terms of remote monitoring, greater device connection results in bigger attack surfaces. This necessitates sophisticated methods for detecting and correcting security flaws. Requirements extend beyond the technology used in devices and surrounding systems to processes that must be in place in organizations, regulators, and manufacturers.
- *Interoperability*—Systems should be compatible with one another, even if they are offered by different suppliers. Individuals in cardiology, for example, must be transferred between institutions and need careful supervision via ECG, should be attached to different gear during movement due to mismatches.

Aim of Fog Computing in HealthCare

On the basis of the study, there seems to be a connection among medical difficulties, resultant needs, and FC advantages, indicating that FC has the ability to be a catalyst for ubiquitous, interconnected devices in healthcare:

- *Accessibility of Processing Locus*—Whenever scaling, security, and dependability concerns make a cloud-only approach unfeasible, FC can provide the processing capabilities required inside the system to fulfill both legislative and technical requirements. To be effective, such approaches require not only the availability of computer resources between sensors and the cloud, but also their efficient management. This allows for program execution visibility as well as flexibility in where processing may take place. The location of fog computing can be dynamic and determined by the modern scenario, surroundings, and application requirements.
- *Integration*—The deployment of novel sensing equipment in today's world usually demands the parallel creation of an architecture. For example, the heart rate tracking system is an instance of a technology that requires specialized equipment. When it relates to producing new, creative technology, this is a huge burden. New devices can be introduced to an established architecture in a FC system. FC may also act as an interconnection layer, translating between different standards.
- *Patient Mobility*—The region where patients may be tracked is likewise limited due to application-specific infrastructure. This is especially important when patients are preparing to depart a hospital's heavily instrumented infrastructure. Existing use scenarios typically fail to handle this shift, which ultimately lengthens a patient's hospital visit. Transitions between various contexts may be managed more progressively with FC.

- *New Applications*—FC allows the creation of whole new applications: Fog computing will minimize delay and response time while also saving energy for ubiquitous and low-cost gadgets executing complex activities such as fall detection [10] by introducing higher degrees of independence and cognition at the edge. Such fog nodes communicate when it is necessary to efficiently manage the network's resources or to control different network-related activities. The intended actions are carried out in a decentralized fashion. Certain challenges include minimizing the overhead email while keeping in mind the safe, private environment in which the IoT devices operate. As multi-hop networks could be untrustworthy, end-to-end protection is required for interactions among fog nodes. To reduce information loss in several areas such as CC, network devices, or IoT devices, privacy is a critical element in CC.

Case Studies

FC can be used in healthcare in various domains. Some of which are explained below.

The Influence of Healthcare 4.0 with FC in Rural Places

Rural communities are underserved in terms of education. The major challenge for high-quality education is the availability of skilled instructors. With the use of emerging innovations like as FC and IoT, educationist offers a mechanism for instructing and studying for learners and lecturers to get real-time tutorials. Furthermore, these techniques assist students in understanding things from the ground up, and these channels of idea exchange assist them in improving their future. These approaches also enable people to address their problems via a common medium and engage with the appropriate individual without delay. Nonetheless, when it comes to medical-related awareness, rural residents suffer greatly owing to a lack of well-trained medical personnel and a lack of knowledge about medical issues. Sometimes, owing to budgetary constraints, the local government and residents of such places might not have access to well-trained medical personnel in their territory. A human being suffers greatly as a result of these difficulties. People in rural regions have historically battled to maintain decent healthcare. The problems that rural hospitals and other institutions confront have been well documented in the literature, and rural Americans are frequently deprived of vital healthcare [25, 26]. Various factors worsened such difficulties in rural regions, including demographic expansion, economic stagnation, a scarcity of medical experts and other medical experts, a huge proportion of elderly, disabled, and unemployed people, and a high incidence of chronic diseases.

Healthcare 4.0 ECG Monitoring [27]

FC can provide unique services to the medical industry, such as support for variety and actual demand fulfillment. It is suitable for analyzing and storing disorganized information. The digital healthcare scanners will be enhanced by dynamically enhancing FC with dispersed storage, combining real-time alert solutions, and data analysis, all of which will be operational at the network's edge. ECG collection is largely used in the treatment of various cardiac diseases. Dynamic time warping (DTW) technique is used to extract information fluctuations from time series and is used in a variety of applications such as walking pattern identification, business, finance, and ECG signals analysis. The primary goal is to characterize arrhythmic ECG bats using QRS-complex measurements and the RR interval. Using real-time signal processing, ECG signals create QRS complexes. This data reduction approach is written in Python and implemented at a fog node. The suggested architecture's ECG signal of 2500 required roughly a moment to assess the fog network. The researchers in [28] proposed the organizational structure seen in Fig. 4.2. To reduce data, the GNU zip software is used to reduce ECG time series. The zipped files are then uploaded to the cloud. The ECG time series sample is displayed by DTW that was broadcast to the Internet. Encoding denotes the rate of ECG compression algorithms. It aids in shortening the time it takes to run the files. ECG data is subsequently handled in smart pathways with the actual separation of heart rate. Figure 4.2 exemplifies the design of an eHealth system based on FC which consists of the three layers. The DL is the first layer, where data from patients is collected via different devices or monitors, such as a smartwatch or smart eyeglasses. All they needed was a link through a Fog gateway and an appropriate communication protocol. The collected data is sent to the fog layer for the preprocessing and after preprocessing, the data is sent to the cloud for storage. The preprocessed data is accessed by the health organizations for treatment.

Patient Training and Monitoring Support [29]

The expense of healthcare and social care for the elderly is rising all around the world. Telecare is thought to compensate for the decrease in conventional clinical contacts and residential care. IoT technologies undoubtedly play a part in conventional telemedicine of chronic illness care, as well as ensuring patients' safety at house and allowing innovative services such as technology aided recovery. Various sensors, both dispersed and mobile, are ideally suited to enhance the security of old and developmentally differently abled patients at home. So far, the typical central server or cloud-based data repository for personal health records (PHRs) has sufficed. The technology and privacy requirements for innovative home telecare systems that incorporate motion capture or continuous activity daily living (ADL) monitoring capabilities are substantially higher, necessitating the adoption of alternative—distributed data processing architectures. The advantages of decentralized fog-like data analysis

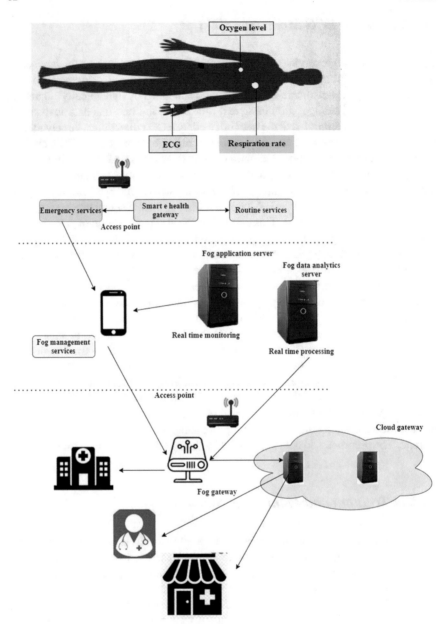

Fig. 4.2 Three layered eHealth architecture

over centralized data analysis for such real-time patient safety surveillance technologies are obvious. Through local data aggregation and decision-making, the distant server load and communication channel throughput may be considerably decreased. Furthermore, from a clinical standpoint, only aggregated meta-information, such as the number of active hours, mean activity level, sleep quality, and presence of exceptional events such as fall downs, have substantial long-term value worth preserving in PHR. The remaining raw data has little clinical use and creates privacy issues if it is transferred outside of private settings. In practice, human accidents could not be detected sufficiently with inertial sensors alone, and dead counting motion monitoring is only reliable for a few meters due to nonlinear effects in IMUs. Due to unforeseen network delays, the specified sensor data fusion cannot be conducted at a remote site. Therefore, sensor data integration, fog-based consolidation, and prospective thinking at the gateway are the ideal solutions for smart household telecare systems that provide ADL analysis and hazard detection. Figure 4.3 depicts everyday living activities and safety monitoring. In the fog layer, sensors collect data and preprocess it. After that, the data is sent to the gateway layer for subsequent operations. Finally, the collected data is sent to the cloud layer for future storage.

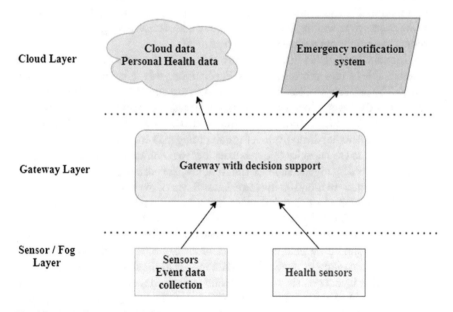

Fig. 4.3 Daily living and activity of data living monitoring

Research Challenges

Fog computing promotes itself as a solution for reducing the complexity of health data administration and, as a result, enhancing its dependability. To that end, before developing a FC-based infrastructure to handle medical information, it is critical to understand the related difficulties which are explained below:

- Management of data: The doctor is presented with a huge quantity of patient information, which is challenged by 5Vs of big data: value, velocity, variety, volume, and veracity. This is primarily concerned with fog nodes (FN's) capacity to take, retain, interpret, and send data. As a consequence, data volatility among the FC and CC might be observed, which should be controlled by the program's management setup. Fog layer (FL) requires common protocols and data formats to manage extremely disparate data such as text, image files, and so on, arriving from various sources like a smartphone or a wristwatch.
- Scalability: Sensors or transportable pharmaceutical products are used to gather information. This program should be provided throughout the facility so that all patients may use their cellphones to obtain healthcare care and get general health information. Furthermore, Healthcare 4.0 may be applied to a full community. It saves clients time waiting for visits and results, as well as giving them immediate access to a minimum degree of health treatments.
- Security and heterogeneous: It is the primary issue, and end-user data should not be accessible by unauthorized entities. Furthermore, this endangers one's personal safety. The primary problem in Healthcare 4.0 implementation and deployment is security and privacy. However, protection is necessary in all levels, including fog layer (FL), device layer (DL), and cloud layer (CL), as well as the unification of these layers. Another difficulty that fog computing may encounter is heterogeneity, which refers to the capacity of diverse devices to communicate with one another. Multiple diverse gadgets, such as smartphones, self-driving vehicles, and other IoT smart items, are found on the lowest most layer of the IoT paradigm. The issue with heterogeneity in IoT end devices includes data gathering, data formatting, and capability. Enabling the network to link with a diverse set of sensors to aid in patient monitoring is one instance of how it might be a problem in healthcare. Information should be heterogeneous in order to be transferred to some other device for processing or analysis. As data moves to the next stage, the fog layer appears. This layer consists of compute and network nodes, clusters, routers, and other devices. As a consequence, diversity is becoming an important consideration in architectural design, especially when various measuring devices interface with the patient.

  Because of its location in the network, security in fog computing may be viewed as a research topic. Working at the network edge exposes the risks that do not exist in a well-organized cloud infrastructure. A few of these threats may be a man in the middle attack, where an attacker relays and changes information among two parties. The attack may attack a gateway between a diagnostic imaging sensor and a fog node that analyzes the patient's data in the event of a health service.

If the attacker tampered with the data being examined, it may have significant consequences for the patients' health.

- Standardization, interoperability, and regulations: In Healthcare 4.0, there are no common norms and standards for protocols and interfaces for various goods and equipment. To address this issue, standardization activities, such as a specialized organization to standardize healthcare technology, are necessary. It aids in achieving real-time reaction and resolving data disparities. When standardizing, gadget connections, information collection connectors, routing protocols, and gateway interfaces should all be addressed. Similar to the issue of device diversity, there are several communication protocols available, such as Wi-Fi. As a consequence, in order to be extensible, the FN must translate protocols at multiple internal FL levels, such as core network, messaging layers, and data annotation layers. Furthermore, before healthcare goods are accessible on the market for patients, a regulatory system must be in place.

- Fog computing secure communications: At the fog layer, different strategies for collecting, analyzing, and storing information are delivered, but privacy procedures are not dispersed. IoT implementations should include some level of protection and privacy. Interactions among connected devices are deemed secure in accordance with the IoT communications security procedures in place. When an IoT device wants to offload a processing or storage requirement, it communicates with fog nodes. Any additional actions that occur as part of the network would not be considered part of the fog system. The method of processing and storage specifications released onto fog nodes is not allowed. Even IoT devices must meet the minimum necessary security protocols. Gadgets should interact securely in the setting of an IoT connection.

- Management of resources: When three separate paradigms (IoT, fog, and cloud) are combined inside one system, resource management becomes a key problem. Even when discussing fog computing without the use of the cloud, resource management may be a challenging problem. This is because the fog has fewer compute and storage capabilities than the cloud. In dealing with resources planning involving IoT objects and interaction with the fog layer, the data sent and analyzed among them must have a purpose in order to avoid sharing data redundancy and squandering vital resources. When several users or parts of a system use the same resources, resource planning becomes even more important in order to provide the shortest possible delay to the devices in use.

- Energy minimization: Since fog settings necessitate the deployment of a large number of fog nodes, computing is dispersed and may be less energy-efficient than the centralized cloud model of compute. As a result, reducing energy usage in fog computing is a significant problem.

- Dynamicity: One of the most significant characteristics of IoT devices is their capacity to develop and constantly modify the composition of their operation. The internal characteristics and performance of IoT devices will be altered because of this challenge. Furthermore, portable devices experience software and hardware aging, which results in changes in workflow behavior and device characteristics.

As a consequence, fog nodes will need continuous and smart reorganization of the topological architecture and resources made available to them.

- Storage: Fog nodes have low storage space as compared to cloud computing. As a result, the number of fog nodes may be raised to process the data that takes up a lot of space (image data). Furthermore, the gathering of these data and the processing of the findings adds time overhead.
- Increased cost: The installation of the fog nodes, as well as connected devices among sensor equipment and the cloud layer, increases the overall cost of development. The application of fog computing in healthcare, on the other side, enhances health, saves lives, and increases life span.

Conclusion and Future Work

The benefits of fog computing for smart healthcare include lower response time and low energy usage, variety, and flexibility. The chapter provides details about the architecture, characteristics, and applications of fog Computing. The chapter also provides detailed information about the utilization of FC in healthcare along with the case studies. The research problems encountered when using FC in modern medical systems are also highlighted so that other companies or research groups can further investigate the area. In cloud computing environments, fog computing is a popular trend. Applications are increasingly requiring extensive usage of the cloud. Despite the fact that hardware capacity has risen considerably, health applications require information to be obtained as soon as feasible. This is an area where fog computing is making a significant difference. Filtration, data gathering, and data forecasting are all examples of artificial intelligence solutions that will soon become ubiquitous in clinics. Furthermore, due to greater mobility of equipment and advancements in communication technology, services will be available to outside patient residences, and the health practitioners or medical facility might be located at any time and from any location in the globe. In these situations, FC can help to minimize latency, allowing for real-time and large-scale medical services, with the potential to benefit the world's most poor population, which lacks access to excellent healthcare. It is expected that the aforementioned problems will be addressed soon, and that a higher standard of living and community well-being would be attained with the assistance of FC. The capacity to triage data and make critical judgments inside the device's own context will aid in extracting essential insights from the massive volume of available data. In the future, the implementation of fog computing with healthcare can be done by the use of IoT. The health-monitoring model can be designed with the help of IoT and fog computing to minimize the latency in processing the healthcare data.

References

1. Singh, A., Mahapatra, S.: Network-based applications of multimedia big data computing in IoT environment. Intell. Syst. Ref. Libr. **163**, 435–452 (2020)
2. Katal, A., Sethi, V., Lamba, S., Choudhury, T.: Fog computing: issues, challenges and tools. In: Emerging Technologies in Data Mining and Information Security, pp. 971–982 (2021). https://doi.org/10.1007/978-981-15-9927-9_92
3. Topol, E.: Creative destruction of medicine: how the digital revolution and personalized medicine will create better healthcare. Hir **19**, 229–231 (2012)
4. Mahapatra, S., Singh, A.: Application of IoT-based smart devices in healthcare using fog computing. In: Studies in Big Data, pp. 263–278 (2020). https://doi.org/10.1007/978-981-15-6044-6_11
5. Wearable Healthcare Technology and Devices|MC10. https://www.mc10inc.com/
6. Kang, D.Y., et al.: Scalable microfabrication procedures for adhesive-integrated flexible and stretchable electronic sensors. Sensors **15**, 23459–23476 (2015)
7. Bonomi, F., Milito, R., Zhu, J., Addepalli, S.: Fog computing and its role in the Internet of Things. In: Proceedings of the First Edition of the MCC Workshop on Mobile Cloud Computing—MCC '12 (2012). https://doi.org/10.1145/2342509
8. Aazam, M., Huh, E.N.: Fog computing: the cloud-IoT/IoE middleware paradigm. IEEE Potentials **35**, 40–44 (2016)
9. Masip-Bruin, X., Marín-Tordera, E., Tashakor, G., Jukan, A., Ren, G.J.: Foggy clouds and cloudy fogs: a real need for coordinated management of fog-to-cloud computing systems. IEEE Wirel. Commun. **23**, 120–128 (2016)
10. Cao, Y., Chen, S., Hou, P., Brown, D.: FAST: a fog computing assisted distributed analytics system to monitor fall for stroke mitigation. In: Proceedings of the 2015 IEEE International Conference on Networking, Architecture and Storage, NAS, pp. 2–11 (2015). https://doi.org/10.1109/NAS.2015.7255196
11. Ai, Y., Peng, M., Zhang, K.: Edge computing technologies for Internet of Things: a primer. Digit. Commun. Netw. **4**, 77–86 (2018)
12. Yi, S., Li, C., Li, Q.: A survey of fog computing: concepts, applications and issues. In: Proceedings of the International Symposium on Mobile Ad Hoc Networking and Computing (MobiHoc) vol. 2015-June, pp. 37–42 (2015)
13. More, P.: Review of implementing fog computing. IJRET: Int. J. Res. Eng. Technol. **4**, 2319 (2015)
14. Bertini, M., Marcantoni, L., Toselli, T., Ferrari, R.: Remote monitoring of implantable devices: should we continue to ignore it? Int. J. Cardiol. **202**, 368–377 (2016)
15. Transforming health: shifting from reactive to proactive and predictive care—MaRS Discovery District. https://www.marsdd.com/news/transforming-health-shifting-from-reactive-to-proactive-and-predictive-care/
16. Rahmani, A.M., et al.: Exploiting smart e-Health gateways at the edge of healthcare Internet-of-Things: a fog computing approach. Futur. Gener. Comput. Syst. **78**, 641–658 (2018)
17. Thota, C., Sundarasekar, R., Manogaran, G., Varatharajan, R., Priyan, M.K.: Centralized fog computing security platform for IoT and cloud in healthcare system. In: Fog Computing: Breakthroughs in Research and Practice, pp. 365–378 (2018). https://doi.org/10.4018/978-1-5225-5649-7.CH018
18. Mutlag, A.A., et al.: MAFC: multi-agent fog computing model for healthcare critical tasks management. Sensors (Switzerland) **20** (2020)
19. The Eight Principles of Patient-Centered Care—Oneview Healthcare. https://www.oneviewhealthcare.com/blog/the-eight-principles-of-patient-centered-care/
20. Farahani, B., et al.: Towards fog-driven IoT eHealth: promises and challenges of IoT in medicine and healthcare. Futur. Gener. Comput. Syst. **78**, 659–676 (2018)
21. Iorga, M., et al.: Fog computing conceptual model. (2018).https://doi.org/10.6028/NIST.SP.500-325

22. Eide, R.B.: Low energy wireless ECG—an exploration of wireless electrocardiography and the utilization of low energy sensors for clinical ambulatory patient monitoring. Medicine (2016)
23. Paksuniemi, M, Sorvoja, H., Alasaarela, E., Myllyla, R.: Wireless sensor and data transmission needs and technologies for patient monitoring in the operating room and intensive care unit. In: Conference Proceedings : Annual International Conference of the IEEE Engineering in Medicine and Biology Society, vol. 2005, pp. 5182–5185 (2005)
24. Alesanco, Á., García, J.: Clinical assessment of wireless ECG transmission in real-time cardiac telemonitoring. IEEE Trans. Inf Technol. Biomed. **14**, 1144–1152 (2010)
25. Ermann, D.A.: Rural healthcare: the future of the. Hospital (2016). https://doi.org/10.1177/107 75587900470010447,33-73
26. Xu, Q., Ren, P., Song, H., Du, Q.: Security enhancement for IoT communications exposed to eavesdroppers with uncertain locations. IEEE Access **4**, 2840–2853 (2016)
27. Jain, R., Gupta, M., Nayyar, A., Sharma, N.: Adoption of fog computing in Healthcare 4.0. In: Signals and Communication Technology 3–36 (2021). https://doi.org/10.1007/978-3-030-461 97-3_1
28. Kumari, A., Tanwar, S., Tyagi, S., Kumar, N.: Fog computing for Healthcare 4.0 environment: opportunities and challenges. Comput. Electr. Eng.**72**, 1–13 (2018)
29. Tammemäe, K., Jantsch, A., Kuusik, A., Preden, J.S., Õunapuu, E.: Self-aware fog computing in private and secure spheres. In: Fog Computing in the Internet of Things: Intelligence at the Edge, pp. 71–99 (2017). https://doi.org/10.1007/978-3-319-57639-8_5

Chapter 5
Intelligent Self-tuning Control Design for Wastewater Treatment Plant Based on PID and Model Predictive Methods

Ujjwal, Neelam Verma, and Anjali Jain

Introduction

Intelligent tuning of a controller is most important function in control engineering field in obtaining the most optimal and efficient parameters and as a result accomplishing the most magnificent system's response [1]. Various techniques have been tested and implemented across the wide spectrum of engineering applications, but it is still the research interest for upcoming technologies. Multivalued and multi-stage plant are having very complex model and highly nonlinear and unstable behaviour and because of that search of optimal and energy efficient control techniques in this domain is highly desired. Since parameters and external disturbance are highly unpredictive and dominating in real-time operation of these plant, automatic tuning is desired for satisfactory plant operation. In the present work, dynamic plant considered for testing effectiveness of suggested control techniques is wastewater treatment plant. Activated sludge process model is energy efficiently controlled as it is a biological wastewater treatment process. First, decentralized PI controller has been implemented which is self-tuned and is adaptable to change the controlled variable of the process for a multivariable system [1]. This controller has been proposed with an adaptive interaction algorithm and examined to the biological wastewater treatment plant. For this, an interaction method named as relative gain array is used to ensuring the proper pairing in between the input and output variables of the plant [2]. Second control technique is PID controller with well-tuned parameters which has been designed with activated sludge process model [3]. In this, two PID controllers

Ujjwal · N. Verma (✉) · A. Jain (✉)
Department of Electrical and Electronics Engineering, Amity University Uttar Pradesh, Noida, India
e-mail: neelamverma11@gmail.com

A. Jain
e-mail: anjalijain.121@gmail.com

are used with given dissolve oxygen and substrate concentration as reference values. Next is model predictive controller which is based on the future predicted values and current measurement. It is an advanced control technique [4, 5]. It is mostly used controller when in control system. At last, Simulink model with different control techniques has been designed with proper pairing of the input–output, and all the simulation results of the plant have been plotted using MATLAB Simulink with given constant reference values and conclude the work based on the results. All controllers have been given satisfactory results, but MPC gave better results as compared to other control techniques. The main aim of this work is to find the best possible control structure, mathematical modelling of multivalued nonlinear plant, designing of PID and model predictive control for considered system, Simulink model with different control techniques has been designed, comparison of PID and model predictive control results, and model predictive control gives improvement performance.

This paper has following five parts. Section "Introduction" gives a brief introduction of the paper. Section "Modelling of Wastewater Treatment Plant" gives the modelling of the wastewater treatment plant. Section "Control Techniques" gives the idea of various control techniques used in this paper. Section "Simulation Result and Discussion" represents the results of the nonlinear system. Section "Conclusion" concludes the results obtained in this paper.

Literature Review

A decentralized PI controller has been used with wastewater treatment plant which is a nonlinear multivariable plant [1]. Also, an interaction method is used for proper pairing of input and output of the plant. PID controller has been designed and then applied to the activated sludge process for removal of ammonium and removal of nitrate [2]. This controller has some advantage over other controllers. A new idea of implementation of PID controller with some reduction methods using particle swarm optimization technique has been implemented for tuning of controller parameters in [4]. Here, self-tuned PI settled very fast and it easily followed the given reference value and has less.

Four advanced control techniques have been implemented, and their comparison is discussed in [4]. Implementation of the advanced approach model predictive control with activated sludge process model with simulation results of dissolve oxygen and substrate concentration has been given in [5]. In paper [6], control technique PI with MPC has been implemented and obtained the simulation results with MATLAB Simulink. Paper [7] shows the work based on the modelling of the wastewater treatment plant. In this, modelling of the given plant has been studied, and control techniques have been introduced.

In the paper [8], performance of wastewater plant has been obtained with PID controlled sludge process with consideration of higher order of the harmonics. Robust performance and stability of the system have been obtained even when noise was present in the system. Paper [9], MPC with PID controller in real-time process control system has been implemented for real-time operation. Paper [10] shows the effect of interaction methods for the multivariable system with the comparison between these interaction methods. Paper [11] discusses the controller tuning based on multiple variables of a system and stability of proposed method.

Objective of the Work

The main aim of this research is to design self-tuning control method to reduce the problem for wastewater treatment plant. Because the WWTPs are highly affected in presence of disturbances and suppressed substance which is related to the incoming wastewater composition. It is determined that the preferred manipulate variable and the set of rules which is tuned were correctly way out the plant output is applied with the activated sludge process model, and the mathematical equations are also proposed. In this way, the PI decentralized control method and tuning methods for several control system and for the system which is nonlinear and have many variables. Here, conventional PID controller and advanced method model predictive control (MPC) have been applied to the plant model.

This work provides to control of the concentration of dissolved oxygen during aeration in wastewater plant and provides the control of the concentration of substrate. Main objective of this research is to design some control techniques for wastewater treatment plant. Also, other works are as follow:

- To study the biological model in the wastewater treatment plant.
- To model the activated sludge process model.
- To design some control techniques which help in reducing the organic contents and improve the effluent water quality.
- To remove the stricter demand of effluent to optimize the nitrate removal.
- To observe the results of the dissolve oxygen and substrate with these control techniques.
- Compare the simulation results obtained from these control techniques.

Proposed Control Methodology

Designing of control system of dynamic plant starts with mathematical modelling with consideration of all variable/parameters of the system, interference and noise, external parametric variations, etc., and then control techniques should be implemented. Here is the proposed approach of work (Fig. 5.1).

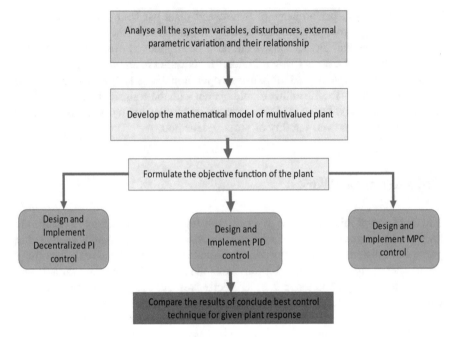

Fig. 5.1 Proposed methodology flow chart

Modelling of Wastewater Treatment Plant

This section includes the brief introduction of nonlinear wastewater treatment plant. Modelling of activated sludge process is very common to design and performance in this plant.

Wastewater Treatment Plant

Wastewater is water that has been polluted to the extent that it is no longer advantageous, before it tends to be utilized back into the earth. In this plant, Fig. 5.2 has three stages which are explained as.

Primary Treatment

It is all about separation of suspended solids from the water. It includes some methods named as bar screens and grit chamber, sedimentation tank, and chlorination of effluent.

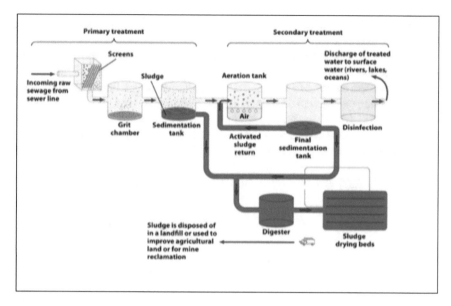

Fig. 5.2 Stages of wastewater plant [12]

Secondary Treatment

It is a biological treatment which includes like activated sludge, trickling filter, and oxidation ponds. In activated sludge process, both aerobic and anaerobic bacteria exist [11]. It promotes the bacteria growth and decompose the organic matter [13]. From this treatment, the trickling filters finally satisfy the condition of the removal of sludge. Finally, clarifiers are used for settling the sludge. And from these, the water is flow back to the rivers.

Tertiary Treatment

In secondary treatment, plant still contains some pathogenic bacteria and nutrients like nitrogen and phosphorus. Due to these matters, water requires tertiary treatment. This stage contains two processes named as nitrification and denitrification [13].

Activated Sludge Process

The activated sludge process is the widely used concept in this process which consists of microorganism which are oxidize and produce the organic materials [9]. Then, the obtained organic matter is converted into carbon dioxide and remaining form in the new cell mass. In the cell mass, there is sludge present which consists living and dead organisms [12].

Usually, this plant is to remove organic matter and suspended substance before it reaches to the recipients. Figure 5.3 represents the bioreactor that describes the biological process contains the aeration tank and clarifier settler. In first tank, microorganism has been oxidized to organic materials. The biomass in the system is present in the settler which has been used to produce the fine quality effluent water. Then, the biomass has been recycled back to the aeration tank. The mass balance components are used to represent the dynamic modelling of the activated sludge process [1, 11].

$$\dot{X}(t) = \mu(t)X(t) - D(t)(1+r)X(t) + rD(t)X_r(t) \tag{5.1}$$

$$\dot{S}(t) = -\frac{\mu(t)}{Y}X(t) - D(t)(1+r)S(t) + D(t)S_{in} \tag{5.2}$$

$$DO(t) = \frac{K_o\mu(t)X(t)}{Y}D(t)(1+r)DO(t) \\ + K_{la}(DO_sDO(t) + D(t)DO_{in}) \tag{5.3}$$

$$\dot{X}_r(t) = D(t)(1+r)X(t) - D(t)(\beta+r)X_r(t) \tag{5.4}$$

where

$X(t)$	concentration of biomass,
$S(t)$	concentration of substrate,
$DO(t)$	concentration of dissolve oxygen,
$X_r(t)$	concentration of recycles water,
$D(t)$	rate of dilution,

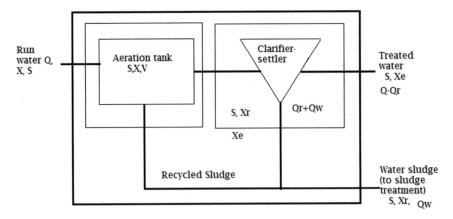

Fig. 5.3 Schematic diagram of activated sludge process [14]

$S_{in}(t)$ substrate of influent water,
$DO_{in}(t)$ dissolve oxygen of influent water, and
$DO_s(t)$ maximum DO concentration.

In this paper, we have two manipulated inputs such as rate of dilution ($D(t)$) and aeration rate ($W(t)$) and two controlled outputs named as substrate ($S(t)$) and dissolve oxygen ($DO(t)$) for simulation in MATLAB Simulink [15].

Control Techniques

In this chapter, the control techniques which have been proposed to the wastewater treatment plant are described. These control techniques are decentralized PI controller with an adaptation algorithm, PID controller with well-tuned parameters, and model predictive control has been implemented to the subsystem [8].

Decentralized PI Controller

It is one of the simpler approaches for the designing of a multivariable controller. In multivariable system, if any input change, then almost all the outputs because of the interact of the input and outputs variables [7]. This controller functions admirably if $G(s)$ is a diagonal, subsequently the controlled plant is basically a group of autonomous plant, and every component in $G(s)$ might be developed freely. The configuration can be a SISO relying upon chosen proper pairing. The equation of PI controller is shown below. This can be emphasized that the error signal $e(t)$ uses in the generation of different action, also with subsequent signs and added with the control signal $u(t)$ which has been applied to the plant.

$$u(t) = K_p e(t) + K_i \int e(t)dt \qquad (5.5)$$

where

K_p is proportional action,
K_i is integral action,

In the form of transfer function, Eq. (5.5) can be written as

$$G(s) = \frac{U(s)}{E(s)} = K_p\left[1 + \frac{1}{T_i s}\right] \qquad (5.6)$$

Here, T_i is known as integral time, and $G(s)$ is a plant that is to be managed with the help of diagonal controller.

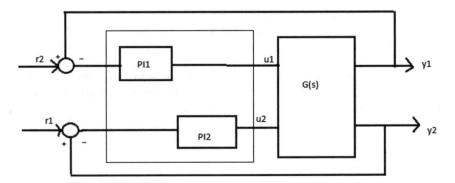

Fig. 5.4 TITO plant with diagonal control [1]

In this work, a TITO, i.e. wastewater treatment plant system, is considered. For this, the given plant is created with two subsystems with each input/output pairs, such as (u_1, y_1) and (u_2, y_2) are controlled with the help of PI1 and PI2, respectively. The parameters of PI to be adjusted in obtaining good dynamic response of the plant which must be controlled. Figure 5.4 shows the diagonal control of TITO system [14].

In this figure, $r1$ and $r2$ are the reference inputs where the feedback of the outputs is added. The inputs of the system are represented by $u1$ and $u2$ applied to the plant and produce the outputs. Here, PI1 and PI2 are two controllers.

PID Controller

PID is proportional integral derivative [3]. It is frequently used controller in control system area due to its efficiency and reliability. It is a controller with feedback which is used for many control problems. In this, there are some set points and some measured variables and output, i.e. control variable, using them, it finds the errors (e). PID controller is commonly used controller in industrial plant [4]. It directly measures the reference value and the output of the plant and then finds the error and reduces the error so that it can reach the set value. The equation of this controller is given as

$$u = K_p e(t) + K_i \int e(t) dt + K_d \frac{de}{dt} \qquad (5.7)$$

Here, K_p, K_i, and K_d are the parameters of PID controller.

Each parameter of the PID represents some particular action based on error: proportional action, integral action, and derivative action.

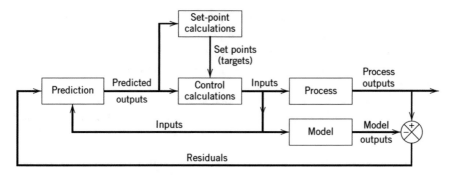

Fig. 5.5 Model predictive control [18]

Model Predictive Control

It is most popularly used controller which is based on the current measurement and future values. Also, it is an optimization technique which is based on a process model which helps to predict the current input values for the actual output. Then compare the actual output with the future values which are obtained form process model [8]. Based on the difference between the values, it proceeds to the predication block and then tries to help in solving the optimal control system problem with a finite horizon at each sampling time [11, 15]. MPC has three functional blocks, named as state estimation, optimizer, and a plant or process model. Optimizer helps to find the optimal control signal form model and minimize the objective function after giving to these values to the plant. Plant model is nothing but predicts the future values for the actual output [7]. Based on predictive output, there are two calculations named as set point calculation and control calculation based on the set points Which are done, as shown in Fig. 5.5 [14]. The MPC is used where the problem exists with such controller such as PID like large time delay, i.e. when response takes long time and where input and output violations exist. It is also used for multi-input and multi output system [16, 17].

This control technique is required to obtain the optimal control signals from which the response of predicted values to reach its target which is fixed values at every time instant.

Simulation Result and Discussion

In this part, first, a subsystem of the activated sludge proceed has been created in MATLAB Simulink. In WWTP, the controlled output variable is DO–dissolved oxygen and S-substrate concentration. The manipulated variables are D-dilution rate and W-aeration rate. ASP model is modelled based on the nonlinear differential

equations described in Sect. "Modelling of Wastewater Treatment Plant" [19, 20].
After this, control techniques have been implemented.

First, a decentralized PI controller has been proposed with reference constant
input, and initial condition for dissolve oxygen concentration and substrate concentration is set as 41 (mg/l) and 2 (mg/l), respectively.

After running the simulations, results obtained are showcased as under.

In Figs. 5.6 and 5.7, it can be seen that the substrate concentration and dissolve
oxygen with respect to time easily follow the reference value and reach to their steady
state.

After that, a conventional PID controller has been implemented to the ASP model.
There 2 PID controllers have been used. One for controlling the concentration of
substrate, and another is for dissolve oxygen concentration with some fixed input
values. The simulation results for this have been shown in Figs. 5.8 and 5.9 as under.

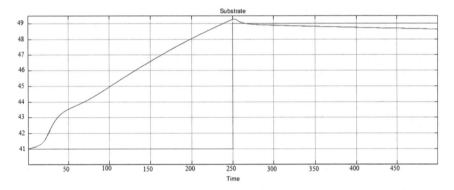

Fig. 5.6 Simulation of substrate in ASP

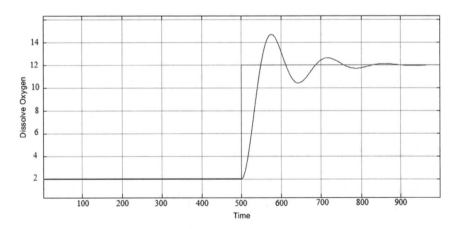

Fig. 5.7 Simulation of dissolve oxygen (mg/l) w.r.t time

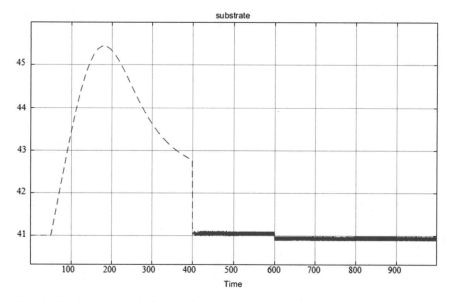

Fig. 5.8 Simulation result of substrate of ASP with PID

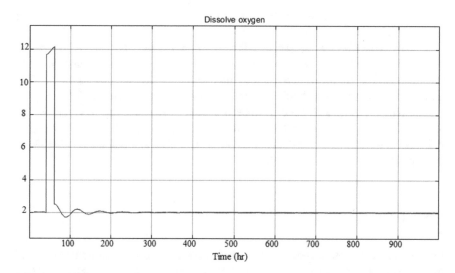

Fig. 5.9 Simulation result of dissolve oxygen of ASP with PID

In Fig. 5.8, result of substrate has been obtained. It increases and after reaching at peak point starts decreasing but it takes time to settle down and finally reaches to its steady state with some minor oscillation. Figure 5.9 shows the simulation result of dissolved oxygen of ASP with PID. Dissolve oxygen concentration has also some oscillation.

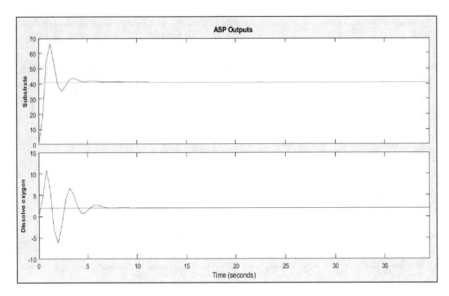

Fig. 5.10 ASP outputs with MPC

Now after implementing, ASP with decentralized PI controller and PID controller MPC will also implemented.

For ASP with MPC, we must design this controller in MATLAB. In this, the control horizon is defined as 10 and prediction horizon as 7 with sampling time which is 0.04 s. And then convert it into discrete time domain. And then plot its output. The outputs for ASP model with this control techniques are shown in Fig. 5.10.

The output results with MPC have been obtained. The substrate has taken 8.73 s to settle down with minor oscillations. And in dissolve oxygen, it easily followed its reference path with two oscillations, but after time 8.38 s, it started to settle down and finally reached to its steady state.

Conclusion

Intelligently, tune optimal controlling techniques are of highest important in the field of multivalued pant as it consumes huge amount of energy for its operation. This work demonstrated the techniques of PID and model predictive control (MPC) application for considered multivariable plant [18]. It has been found that the preferred control techniques which are intelligent-auto tuned have been tracked effectively the reference points for simple and nonlinear plants with impressive consequences for the tuning of the system. The simulation results of that systems have been obtained and plotted using MATLAB Simulink as discussed in the previous sections. In this paper, the MPC control technique provides an optimal and energy efficient result of the system which is nonlinear and has many variables. The concentration of substrate and dissolve oxygen has been tracked the reference path very efficiently and effectively. These two take less time to reach its stable state as compared to other two control technique. And output results with MPC have been obtained with one or two oscillation and overshoot. So, a nonlinear multi-input and multi-output WWTP have been controlled with the help of PID and model predictive control methods for the activated sludge process model.

Future Scope

We must concern about removal of all inorganic materials which are harmful for us and our environment. And some day, today's issues with wastewater will be solved and that water will become drinkable. We should focus on how we can reuse water, remove micropollutants and pathogens, recycle nutrients produce the energy, and reuse the sludge with its resources. The future scope for the same wastewater treatment plant some advanced control techniques can be implemented for the better performance.

Future scope of work also includes the cost-effective management of energy used in wastewater treatment plant along with minimizing the volume and quantity of solid generation without compromising the quality of treated water. Also, such plants should be made self-sufficient to deal with energy required for treatment. The enhanced research needs to be carried out for nutrient (such as ammonia, nitrogen, and phosphate) mining and recovery, and reuse of biosolids.

Also, in this pandemic era, the future research includes preparedness of wastewater treatment plants to deal with new strains of viruses and bacteria, or any spill incidents.

References

1. Samsudin, S.I., Rahmat, M.F., Wahab, N.A., Razali, M.C., Gaya, M.S., Salim, S.N.S.: A self-tuning decentralized PI control for multivariable plant. In: IEEE International Conference on Control System, Computing and Engineering, Penang, Malaysia, 23–25 Nov 2012
2. Yadav, A.K., et al.: Soft computing in condition monitoring and diagnostics of electrical and mechanical systems. In: 1E Part of the Advances in Intelligent Systems and Computing, vol. 1096, pp. 496. Springer Nature, 2020, ISBN 978-981-15-1532-3. https://doi.org/10.1007/978-981-15-1532-3
3. Kodaly, R.K.: Smart wastewater treatment. In: 2017 IEEE Region 10 Symposium (TENSYMP)
4. Blondin, M.J.: Optimization algorithms in control systems. In: Controller Tuning Optimization Methods for Multi-Constraints and Nonlinear Systems, pp. 1–9. Springer, Cham (2021)
5. Choo, H.P., Sablan, S., Eek, R.T.P., Wahab, N.A.: Self-tuning PID controller Co for activated sludge system. In: IEEE 8th Conference on Industrial Electronics and Applications (ICIEA) (2013)
6. Rahmat, M.F., Samsudin, S.I.: Control strategies of wastewater treatment plants. Aust. J. Basic Appl. Sci. **5**(8), 446–455 (2011)

7. Holenda, B., Domokosa, E., Redey, A., Fazakas, J.: Dissolved oxygen control of the activated sludge wastewater treatment process using model predictive control. Comput. Chem. Eng. **32**, 1270–1278 (2008)
8. Harja, G., Nascu, I.: MPC advanced control of dissolved oxygen in an activated sludge wastewater treatment plant. www.researchgate.net/publication/305674235
9. Srivastava, A., Prasad, L.B.: A comparative performance analysis of decentralized PI and model predictive control techniques for liquid level process system. Int. J. Dyn. Control 1–12 (2021)
10. Hamad, D., Dhib, R., Mehrvar, M.: Identification and model predictive control (MPC) of aqueous polyvinyl alcohol degradation in UV/H 2 O 2 photochemical reactors. J. Polym. Environ. 1–13 (2021)
11. Norhaliza A. Wahab, Reza Katebi, Jonas Balderud, "Multivariable PID control design for activated sludge process with nitrification and denitrification", Biochemical Engineering Journal Elsevier, vol.45, pp. 239–248, 2009
12. Cho, J.-H., Sung, S.W., Lee, I.-B.: Cascade control strategy for external carbon dosage in predenitrifying process. Water Sci. Technol. **45**(4–5), 53–60 (2002)
13. Srivastava, S., et al.: Applications of artificial intelligence techniques in engineering—volume 1. In: 1E Part of the Advances in Intelligent Systems and Computing, vol. 698, 643p. Springer Nature (2018). ISBN 978-981-13-1819-1. https://doi.org/10.1007/978-981-13-1819-1
14. Nikita, S., Lee, M.O.: Control of a wastewater treatment plant using relay auto-tuning. Korean J. Chem. Eng. **36**(4), 505–512 (2019)
15. Bristol, E.: On a new measure of interaction for multivariable process control. IEEE Trans. Autom. Control **76**, 133–134 (1965)
16. Schnobrich, M.R., Chaplin, B.P., Semmens, M.J., Novak, P.J.: Stimulating hydrogenotrophic denitrification in simulated groundwater containing high dissolved oxygen and nitrate concentrations. Water Res. **41**(9), 1869–1876
17. Flanagan, M.J., Bracken, B.D., Ressler, J.F.: Automatic dissolved oxygen control. J. Environ. Eng. Div. **103**(4), 707–722 (1977)
18. Rahman, M.F., Shamsuddin, S.I., Wahab, N.A., Salim, S.N.S., Gaya, M.S.: Control strategies of wastewater treatment plant. Aust. J. Basic Appl. Sci. **5**(8), 446–455 (2011) IISSN 1991-8178
19. Srivastava, S., et al.: Applications of artificial intelligence techniques in engineering—volume 2. In: 1E Part of the Advances in Intelligent Systems and Computing, vol. 697, 647p. Springer Nature (2018). ISBN 978-981-13-1822-1. https://doi.org/10.1007/978-981-13-1822-1
20. Halverson, B.: Interaction analysis in multivariable control systems. Uppsala Dissertations from the Faculty of Science and Technology vol. 92, 162pp. Uppsala. ISBN 978-91-554-7781-3

Chapter 6
Impact of Deep Learning Models for Technology Sustainability in Tourism Using Big Data Analytics

Ashish Kumar and Rubeena Vohra

Introduction

Tourism industry is a service industry that demands accurate forecasting of information related to tourist planner, demand forecasting, personalized travel recommendation, pricing strategies, and many more related services. To improve travel services for customer satisfaction and experience, there is a need to analyze the available structured and non-structured data. For future tourism enhancement, there is a need of precise analysis of available big data for tourist planner, sentiment analysis, opinion analysis, and emotion analysis [1, 2]. For this, deep learning is considered to be as the technology support which not only improves the future of tourism industry but also provides facility to customers in terms of pricing comparison, accommodation booking, and competitive analysis.

Deep learning-based methods in the tourism industry provide automation in terms of booking flow, tourist planner, pricing, and business strategies. Tourism models exploiting the deep learning techniques such as Recurrent Neural Network (RNN), Long Short-Term Memory (LSTM), and character-level Convolutional Neural Network (CharCNN) have been proposed for forecasting tourist demand [2], tourist flow [3], and textual sentiment analysis [4]. Figure 6.1 illustrates about the tourism model architecture. The various applications of deep learning models in the field of tourism has been divided into three levels. Initially, the sources of input information are explored. Then, the deep learning technologies which are exploited

Present Address:

A. Kumar (✉)
School of Computer Science, Engineering and Technology, Bennett University, Uttar Pradesh Greater Noida, India
e-mail: ashish.gupta14d@gmail.com

R. Vohra
Bharati Vidyapeeth's College of Engineering, New Delhi, India

© The Author(s), under exclusive license to Springer Nature Singapore Pte Ltd. 2023
V. Kadyan et al. (eds.), *Deep Learning Technologies for the Sustainable Development Goals*, Advanced Technologies and Societal Change,
https://doi.org/10.1007/978-981-19-5723-9_6

for analysis of tourism model have been detailed. These models are based on the prior information of the tourist that includes past booking, hotel stay, location explored, and other potential influential attributes. Finally, the obtained informed has been analyzed for improving the tourism models. The analysis based on this information can be utilized by the tourism planner to provide efficient and reasonable services to the customer [5]. Many big travel companies have utilized voice and text recognition system based on deep learning in mobile app and website to improve business. In addition, machine learning techniques namely, Naïve Bayes, Self-organizing Map, Support Vector machine (SVM) are widely utilized for automating tourism related sentiment, personalized recommender system and booking cancelation [6–8].

During Covid era, travel industry is impacted a lot and concern has been raised to overcome the challenges. Due to this booking cancelation is more in comparison to the previous year, worsening the situation as people are reluctant to travel [8]. Customers are not only afraid of Covid-19 infection but also bounded by the travel and quarantine norms imposed by the government [9]. In order to sustain in this pandemic, travel industry brands need to explore new innovative ways to attract the customers. Industry demands to evolve out of this situation by considering the travel behavior

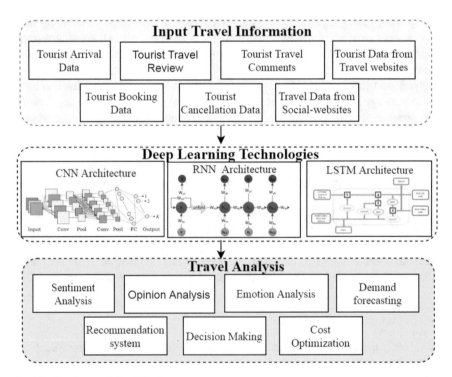

Fig. 6.1 Tourism model architecture

and restrictions. In addition, industry needs some recent and advanced models based on deep learning technique to predict innovative ways to attract customers to enhance business. Deep learning-based tourism models can analyze large amount of online data namely, customer comments, reviews, and recommendations. These models can help in prediction of certain possible outcome which can remain unnoticed by other technology. In sum, deep learning can provide enhancement to tourism industry by implementing analytical models based on available big online data. This can help both the traveler and the industry by providing the hybrid ways to deal with current dynamic customer requirements and the everchanging government regulations in terms of Covid-19 pandemic.

With the progress in technology, demand of big data analytics in tourism has been increased. The data can be collected either from the environment or from the customer itself. The environment includes the destination geo location information, location photos, climate details, and other seasonal information [10]. The customer itself utilized the either from its previous travel or the comments and reviews from the travelers who have experienced the services. In other words, information can be gathered either offline or online. With the change in travelers pattern and requirements, online information has become the foremost source of recommendation to the customers. With the available option of online ticket booking, hotel booking, booking cancelation, and other benefits, the big data analytics in tourism has been substantially to attract the customers.

In this chapter, we will be exploring the utilization of deep learning-based method in the tourism domain. We will be reviewing the existing solutions and analyze their strengths and weaknesses. We will be checking the innovative ideas that can help the travel industry to improve business by attracting customers and optimizing cost. Also, the existing work can be analyzed to provide the effective, efficient and reasonable solution to revamp the industry in the present scenario.

Deep Learning Progress in Tourism

In the recent year, deep learning-based tourism models have gained momentum in the industry. Deep learning based models offers insightful and innovative suggestions that are helpful in the growth of the industry and provide benefits to the customers. In this direction, tourist demand forecasting model were proposed exploiting the artificial neural network architecture such as multilayer perceptron [11], radial basis function neural network [11], and denoised neural network [12]. In Claveria et al. [11], authors exploited a single hidden layer perceptron model with optimum number of neurons. In addition, radial basis function neural network was also implemented on the same dataset. Silva et al. [12] proposed denoising neural network for predicting tourist seasonal demand at European country. Further progress in the domain of tourist demand forecast, many deep learning-based model have been proposed [13, 14]. These models have utilized the ensemble deep learning, GRU, LSTM, and BiLSTM for forecasting tourist demand. Similarly, for sentiment analysis recent

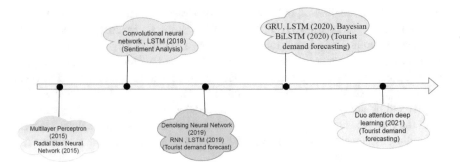

Fig. 6.2 Progress of deep learning methods in various tourism models

models exploiting Convolutional Neural Network, Long Short-Term Memory has been proposed [15–18]. Deep learning-based Chatbot was proposed to enhance tourist culture experience during journey [19]. Figure 6.2 illustrates the progress of deep learning methods in various tourism models in the past few years.

With the progress of deep learning in tourism, the industry has been transformed digitally to offer serve the customer with advanced resources and application. Face recognition system based on deep learning have reduced the requirement the paper bound process. For immigration check, customer identity check, and other security related process have been streamlined with the innovation in the field. The customer reviews and sentiments from social media and traveler website have been analyzed to provide better recommendation to the customers. These models can reduce the time and efforts of the customers and provide them the options which cannot be possible with human analysis. In order to facilitate the tourist in a better way, tourist demand forecasting system were designed with the help of deep learning. These model help the tourist planners to address resource requirement, booking data and other facilities that need to focus for a set of travelers. Nevertheless, chatbot and conversational apps were also designed to reduce the efforts of the customers. These features understand the customer requirements and provide suggestions based on their specified preferences. In sum, deep learning is an innovation technology that impacted every domain of tourism industry and provides powerful applications. These applications will not only attracted the customers but also, reduces the effort of the travel agencies in terms of time and manpower. In addition, deep learning-based tourism time can help to reduce the expenditure of the travel planners and support the industry to sustain in the Covid-19 situation.

Deep Learning-Based Tourism Models

Deep learning has impacted many real time areas such as computer vision, face recognition, object tracking, and travel industry [12, 20]. In travel domain, deep learning can primarily attain certain tasks that include identification of best possible candidate for conversion, better analysis of user travel pattern and optimization of cost to improve the present state of the industry. The implication of deep learning in travel domain can provide powerful applications such as chatbot, prediction systems, recommender system, predictive system, demand forecasting, cost management, and sales management. The various tourism model with their specific application areas and the exploited methodology is discussed in turn.

Deep Learning-Based Recommendation System in Tourism

Many AI-based recommender systems have been proposed in travel industry to provide preferences to the customer to advise them for their travel [7]. Generally, the collaborative recommender system are proposed in travel domain using content-based filtering and ensemble learning [5, 7]. In Nilashi et al. [7], authors have proposed travel recommendation system using clustering ensemble methods on TripAdvisor dataset. Hypergraph partitioning algorithm was applied on the clustering results to improve the accuracy of the data obtained. However, Authors have exploited CNN and DNN to propose a personalized recommender system for customers [5]. Textual data in terms of product review and recommendations were analyzed to provide personalized recommendations. However, the model is less effective as no image data was utilized for analysis. Nilashi et al. [14] proposed a recommendation system for recommending eco-friendly hotel based on preference learning. A huge dataset of traveler's rating was clustered to provide the preferences for the hotel booking that includes cost, location, cleanliness, and others parameters. However, the user's online reviews were not considered for the evaluation. However, in Nilashi et al. [15], authors proposed recommendation agents using multi-criteria collaborative filtering using customer's review on socio-networking sites. The methods used for hotel recommendation based on the multi-criterion customer's review. Table 6.1 tabulates the recommendation system proposed in tourism. Apart from recommendation system, predictive systems were also proposed for tourist demand forecasting and sentiment analysis.

Table 6.1 Representative work for recommendation system in tourism

Reference	Technology used	Dataset	Description
Nilanshi et al. [7]	Cluster ensemble	TripAdvisor	Efficient performance metrics used for comparison
Wang [5]	CNN, DNN	Data collected using web crawler	Impact of weather not considered for recommendation
Nilashi et al. [14]	Preference learning	TripAdvisor	Processing traveler's rating for recommending eco-friendly hotels
Nilashi et al. [15]	Collaborative filtering, fuzzy based rule	TripAdvisor	Recommendation agents based on multi-criteria customer's rating

Deep Learning-Based Tourist Demand Forecasting Model

Tourist demand forecasting is very crucial for tourist planners. Demand forecasting model based on deep learning methods facilitate the planners in terms of resource organization, travel management, and booking management [16–18, 21]. In this direction, Zhang et al. [16] proposed a deep learning-based model by analyzing limited data on complex models. This method was helpful for predicting short term decision making and analyzing risk management. The proposed method had shown superior performance in comparison to existing deep learning methods. Authors forecasted tourist arrival data in Beijing by proposing a deep learning tourism model using ensemble learning [17]. For this, authors analyzed the previous tourist and economic data statistics for tourism forecasting. In addition, the work accounted the search intensity index which considered being most representative criteria for accurate estimation of tourist demand. However, in Sun et al. [18] authors proposed a deep learning tourism model exploiting LSTM and GRU for forecasting the tourist arrival in Morocco. The predicted results were more accurate in comparison to machine learning-based model. Kulshrestha et al. [21] proposed a BiLSTM-based deep learning-based tourism model for forecasting the tourism demand of Singapore from five countries. Bayesian network was used for optimizing the hyperparameters. Experimental results proved the superiority of the proposed methods in comparison to other deep learning-based models. In Silva et al. [12], two models namely, parametric and non-parametric were proposed using denoising neural network. The model was applied on the tourist arrival data in European countries. The method was compared with recent model and outperform them statistically. In sum, deep learning-based tourism demand framework can forecast tourism requirement more accurately in comparison to other models. The methods can be adopted by tourist planners to manage their services as per the demand from the tourist. Also, the results

Table 6.2 Representative work for tourist demand forecasting using deep learning

Reference	Technology used	Dataset	Description
Zhang et al. [16]	STL, DADLM	Tourism arrival data from Hong Kong official website	Duo attention layer in deep model to prevent model overfitting
Sun et al. [17]	Ensemble deep learning, Kernel-based extreme learning machines	Tourism arrival data from Beijing official website	Bagging and SAE in deep model to prevent model overfitting
Laaroussi and Guerouate [18]	GRU, LSTM	Tourist arrival data in Morocco	Parameter tuning used to configure model for accuracy
Kulshrestha et al. [21]	Bayesian BiLSTM	Singapore tourism dataset	Tourism demand of Singapore from five different countries considered
Silva et al. [12]	Denoised neural network	European tourist arrival dataset	Statistically determined customer demand with seasonal time series

can help tourist planners to provide a better experience to their customers. Table 6.2 tabulates the various demand forecasting model using deep learning technology. The next section will elaborate about the next application of deep learning in tourism domain.

Deep Learning-Based Sentiment Analysis Models in Tourism Sector

In other line of research in tourism, sentiment analysis and emotion analysis can transform the overall business by providing recommendations to the customers based on people comments and reviews either on social-websites or on tourism websites [22–25]. In this direction, Authors have utilized the customer's review from the social media and sentimentally analyzed using semantic clustering [22]. The proposed method not only exploited the textual information but also the time, geological and weather information to provide quality suggestions to the customers.

The system was more precise in terms of suggestions as multiple contextual parameters were considered for investigating the comment sentiment. In Li et al. [23], authors proposed lexicon based deep learning model integrating CNN-LSTM. The experimental results proved the superiority of the method on Chinese travel reviews. On the other hand, authors exploited visual information in order to analyze the customer behavior and perceptions [24]. Flickr images were used for suggesting destination to the customers for travel. In Li et al. [25], domain specific new words

Table 6.3 Representative work for sentiment analysis in tourism sector

Reference	Technology used	Dataset	Description
Abbasi-Moud et al. [22]	Clustering	TripAdvisor	Multiple factors considered to provide recommendation based on sentiment analysis
Li et al. [23]	CNN-LSTM	Stanford sentiment Treebank	Model computationally complex
Zhang et al. [24]	Deep learning model	Images from Flickr	Destination recommendation based on tourist behavior and perception
Li et al. [25]	Lexicon based model	Reviews from Qunar.com and Ctrip.com	Detect new words in tourism for sentiment analysis
Chang et al. [26]	CNN	TripAdvisor	Considered multiple information about customer to relate to reviews and comments

were proposed to deal with user invented words in tourism domain. Word propagation was used to provide quality sentiment analysis by analyzing sentiment scores with other parameters. Chang et al. [26] proposed a model to analyze hotel reviews, customer's responses and comments using multi-feature and CNN. The study provided the suggestions to customer as well as hotel staff. In sum, a lot of unstructured data in the form of user reviews and comments is available for analysis. Deep learning-based sentiment analysis for tourism models are able to provide better recommendation option to the customer to improve the business in the area. Substantial parameters and details for sentiment analysis model in tourism are tabulated in Table 6.3.

Tourism Sustainability and Covid-19

Tourism sector is widely impacted due to the spread of Covid-19. However, preventive measures must be taken to ensure the sustainability of the tourism sector in the present scenario. During this era of Covid-19, tourism industry has transformed by deploying the deep learning in many areas. Deployments of deep learning in tourism industry ensure to address the tourist requirement in the current situation. In addition, the application must focus on cost optimization in order to reduce the loss that industry faced. For this, industry has incorporated AI-based social media inquire system, automatic food suggestions-based guest reviews and travel apps for reducing the human efforts [27, 28].

Further, to provide customer a satisfying experience, industry should focus on competitive intelligence. In this era, guest services in a hotel such as room cleaning and room delivery can be provided with help of a robot. Voice controlled electronics gadgets can be installed in room to reduce the human–machine physical interaction. Check-in in hotels can be done using facial recognition devices to avoid paper work and physical contact. This can be possible with the implication of deep learning in the domain. This will help the industry to address the challenges that has been modified in many ways. To gain the customer confidence and trust, the industry efforts must be in the direction to overcome the situation demand.

Application to Facilitate Challenges in Tourism

In today's era of Internet, travelers can plan trips, compare options, book a room and cancel a booking through website. With the advancement in technology, now users prefer to do all this stuff through a mobile app which is considered to be easy accessible. However, to provide quick access to services, recommendations and better traveler experience companies are now emphasizing on more advanced and innovative methods. For this, travel companies are also focusing to provide personal recommendations through voice-based interaction systems. In this direction, face expression recognition systems and conversational travel app have been proposed [29, 30]. Conversational app [30] provides personal assistance to customer based on their recent question and history. This saves time and effort as user need not to spend time for reading long documents to take a decision. Chatbots [19] are also used to address the specific query of the customers and saves the cost to the company. Face expression recognition model was proposed to measure customer satisfaction using emotion analysis [29]. Emotion analysis using facial expression provided directions to the companies to improve their service to attract more business.

In recent years to enhance traveler's experience, concept of smart tourism is becoming very popular [31, 32]. Smart tourism utilizes the innovative way of Information Communication Technology (ICT) and artificial intelligence for providing personalized recommendations for smart destination of smart cities. Smart tourism focus to provide quality experience to customers by incorporating social and organizational component together. With the advancement in artificial intelligence, smart tourism relies to provide technological upgradation in terms of business and operations by intelligently processing the data. Integrating recent technologies in tourism provide cross departmental business which help in developing business strategies for smart tourism. This will ensure smart experience to the customer which boom the sector and ensure its steady growth.

In order to address current situation and a boom in the sector, the industry should focus on to provide innovative and new facilities to the customers [33]. The facilities must include the features that involve very less physical movement of the customers. The customer must be provided with easy guidance so that usage of new methodology can be promoted. Online booking and payment should be provided with secure option to generate the trust in the customers. User can book, cancel and postpone a flight booking or hotel booking without any complicated procedure. In addition, customer can also be provided with recommendations based on its past travel and its review to enhance its experience. The methods should also be reviewed to improve marketing and customer retention back in this era.

To summarize, the stakeholders are deploying latest trends in the travel industry to meet the customer's expectation. Voice technology, virtual reality, wearable devices, robot-based assistant, IOT enabled hotel rooms are few recent addition to tourism sector to attract customers. These technological innovations may attract customers who are reluctant to travel as they ensure the government Covid-19 travel policies and restrictions. The high-tech deep learning-based technology can reduce the risk of spread of Covid-19 as human intervention has been restricted in multiple areas.

Impact of Covid-19 on Tourism

In the era of Covid-19, tourism sector is worst hit worldwide due to booking cancelations and refund of cancelation [9, 34]. This had increased the outflow of the money which results in job cutting and severely impact the tourism economy. Government quarantine norms and containment restrictions are the major roadblocks for the fast recovery of tourism ecosystem [35]. In this pandemic tourism industry was first to shut-down and last to restart. Tourism sector can only regain momentum when traveler demand flow will increase. However, the traveler's behavior and confidence has impacted due to ongoing longer pandemic situation. In order to address these issues, we need to employ certain measures to ensure the overall growth of the industry. We need to limit the expenditure of services. However, there should not be any compromise on the quality of services extended to customers. The focus should be on alternate and innovative ideas to ensure the customer satisfaction and requirement which has changed in the current situation.

The pandemic has impacted the Indian tourism as well that lead to economic crisis in the sector [36]. Due to the current situation, count of the international traveler's arrival in India has fallen drastically in 2020. This has caused the decline in foreign exchange, job opportunities and impacted the local vendor associated with the industry. The traditional models proved inefficient to address the current situation effectively. These models had neglected the risk analysis that has come in the form of Covid-19 in the current situation. However, with the advancement in technology, new innovative models will be proposed to provide better insight and future to the industry. With the help of Internet and other digital methods the industry is ready to gain its momentum again [37].

Tourism Sustainability: Post Covid-19

Over the year, tourism is an industry which is impacted either by natural crisis or by manmade crisis. In the year 2020, the industry is not only affected but it has created an economic crisis worldwide [38]. In order to address the current issues and situation, new policies and strategies need to be defined to ensure the fast recovery of the industry. It has been accepted that there is a requirement to deeply analysis the challenges that tourism industry is currently facing. There is a demand of a framework from the market players that can ensure the sustainability of the industry in present situation of Covid-19 [39]. In addition, traveler's planner, tourist demand forecaster, and decision makers need to investigate the imperative areas in the domain to improve the business by attracting customers.

In future, the travel industry should focus on reducing the human efforts and utilize alternative ideas and innovation to ensure customers. Considering the present situation, proper sanitization, hospitality, facilities, and other services can be extended to customers with the help of latest technologies [40]. Post-covid era should focus on more incorporation of technology-based facilities such as robots, chatbot, voice-based app, and other latest inventions. This technological advancement ensures the government restrictions of social distancing and can attract customer confidence in the travel industry. Social distancing has directed the tourism industry to transform the people's travel experience. Travel activities, group activities, and personal activities that travel companies' offers to their customers need to be redesigned in such a way that can gain customer's confidence. During pandemic people interaction, dependency and life style has become more on virtual devices. Their adoption in the domain can improve the customer statistics and enhance the economic ecosystem.

Reduced customer movement, revenue management, and booking cancelation are some negative impacts of Covid-19 on tourism industry. People associated with the industry have lost the employment, motivation, and health [41]. In addition, tourism education has faced reduction in student's intake and research funding. We need to calibrate the new ways of online teaching learning process to attract more students and government funding for the tourism. There is a need of investment in the area to identify the alternative ways to maintain social distancing norms, regain customer confidence, flexible skill set to industry people. In sum, the tourism is facing the physical manpower and revenue crisis due to the ongoing pandemic. To regain the momentum, focus should on the new methodology in synchronization with the existing situation. All the stakeholders, customer, planner, and the supporting staff should be trained and equipped with the latest tools paramount for the quick recovery of the tourism.

Conclusion

In this chapter, we have reviewed and analyzed the recent tourism models exploiting deep learning methods. Various applications of deep learning in tourism domain has been discussed in depth. It has been well acknowledged that deep learning has introduced innovation in tourism industry and utilized methods to improve customer satisfaction and experience. Sentiment analysis, recommendation system, and tourist demand forecast model has provided personalized recommendations and suggestions to customers. These efforts in the domain has introduced the measures which are helpful the sustainability of the industry in the era of Covid-19. Nevertheless, the tourism models are capable to utilize the cost optimization in the area using these methods and can save resources to address the present challenges.

In this study, we have gathered the new innovative trends in the tourism sector which may help in the fast recovery of the industry. The incorporation of the new trends in the domain can decline the cost expenditure. The tourist demand forecast model, hotel recommendation system, voice-based conversational app, chatbot, predictive model based on sentiment analysis and opinion analysis can reduce the efforts of the planners and provide world-class experience to the customer. These measures can increase the revenue in the sector and support in its fast recovery.

References

1. Martín, C.A., Torres, J.M., Aguilar, R.M., Diaz, S.: Using deep learning to predict sentiments: case study in tourism. Complexity **2018** (2018)
2. Law, R., Li, G., Fong, D.K.C., Han, X.: Tourism demand forecasting: a deep learning approach. Ann. Tour. Res. **75**, 410–423 (2019)
3. Zhang, B., Li, N., Shi, F., Law, R.: A deep learning approach for daily tourist flow forecasting with consumer search data. Asia Pac. J. Tourism Res. **25**(3), 323–339 (2020)
4. Kim, Y., Jernite, Y., Sontag, D., Rush, A.M.: Character-aware neural language models. In: Thirtieth AAAI Conference on Artificial Intelligence (2016)
5. Wang, M.: Applying Internet information technology combined with deep learning to tourism collaborative recommendation system. PLoS ONE **15**(12), e0240656 (2020)
6. Kirilenko, A.P., Stepchenkova, S.O., Kim, H., Li, X.: Automated sentiment analysis in tourism: comparison of approaches. J. Travel Res. **57**(8), 1012–1025 (2018)
7. Nilashi, M., Bagherifard, K., Rahmani, M., Rafe, V.: A recommender system for tourism industry using cluster ensemble and prediction machine learning techniques. Comput. Ind. Eng. **109**, 357–368 (2017)
8. Sánchez-Medina, A.J., Eleazar, C.: Using machine learning and big data for efficient forecasting of hotel booking cancellations. Int. J. Hosp. Manag. **89**, 102546 (2020)
9. iLibrary, O.: Tourism Policy Responses to the Coronavirus (COVID-19) (2020)
10. Paolanti, M., Mancini, A., Frontoni, E., Felicetti, A., Marinelli, L., Marcheggiani, E., Pierdicca, R.: Tourism destination management using sentiment analysis and geo-location information: a deep learning approach. Inf. Technol. Tourism **23**(2), 241–264 (2021)
11. Claveria, O., Monte, E., Torra, S.: Tourism demand forecasting with neural network models: different ways of treating information. Int. J. Tour. Res. **17**(5), 492–500 (2015)
12. Silva, E.S., Hassani, H., Heravi, S., Huang, X.: Forecasting tourism demand with denoised neural networks. Ann. Tour. Res. **74**, 134–154 (2019)

13. Srisawatsakul, C., Boontarig, W.: Tourism recommender system using machine learning based on user's public Instagram photos. In: 5th International Conference on Information Technology (InCIT), pp. 276–281. IEEE (2020)
14. Nilashi, M., Ahani, A., Esfahani, M.D., Yadegaridehkordi, E., Samad, S., Ibrahim, O., et al.: Preference learning for eco-friendly hotels recommendation: a multi-criteria collaborative filtering approach. J. Clean. Prod. **215**, 767–783 (2019)
15. Nilashi, M., Ibrahim, O., Yadegaridehkordi, E., Samad, S., Akbari, E., Alizadeh, A.: Travelers decision making using online review in social network sites: a case on TripAdvisor. J. Comput. Sci. **28**, 168–179 (2018)
16. Zhang, Y., Li, G., Muskat, B., Law, R.: Tourism demand forecasting: a decomposed deep learning approach. J. Travel Res. **60**(5), 981–997 (2021)
17. Sun, S., Li, Y., Guo, J.E., Wang, S.: Tourism demand forecasting: an ensemble deep learning approach (2020). arXiv:2002.07964
18. Laaroussi, H., Guerouate, F.: Deep learning framework for forecasting tourism demand. In: IEEE International Conference on Technology Management, Operations and Decisions (ICTMOD), pp. 1–4. IEEE (2020)
19. Sperlí, G.: A Cultural heritage framework using a deep learning based Chatbot for supporting tourist journey. Expert Syst. Appl. 115277 (2021)
20. Kumar, A., Walia, G.S., Sharma, K.: Recent trends in multicue based visual tracking: a review. Expert Syst. Appl. **162**, 113711
21. Kulshrestha, A., Krishnaswamy, V., Sharma, M.: Bayesian BILSTM approach for tourism demand forecasting. Ann. Tour. Res. **83**, 102925 (2020)
22. Abbasi-Moud, Z., Vahdat-Nejad, H., Sadri, J.: Tourism recommendation system based on semantic clustering and sentiment analysis. Expert Syst. Appl. **167**, 114324 (2021)
23. Li, W., Zhu, L., Shi, Y., Guo, K., Cambria, E.: User reviews: sentiment analysis using lexicon integrated two-channel CNN–LSTM family models. Appl. Soft Comput. **94**, 106435 (2020)
24. Zhang, K., Chen, Y., Li, C.: Discovering the tourists' behaviors and perceptions in a tourism destination by analyzing photos' visual content with a computer deep learning model: the case of Beijing. Tour. Manage. **75**, 595–608 (2019)
25. Li, W., Guo, K., Shi, Y., Zhu, L., Zheng, Y.: DWWP: domain-specific new words detection and word propagation system for sentiment analysis in the tourism domain. Knowl.-Based Syst. **146**, 203–214 (2018)
26. Chang, Y.C., Ku, C.H., Chen, C.H.: Using deep learning and visual analytics to explore hotel reviews and responses. Tour. Manage. **80**, 104129 (2020)
27. http://www.travelweekly.com/Travel-News/Airline-News/Artificial-intelligence-driving-KLM-social-media-strategy
28. https://www.buzzfeed.com/josephbernstein/the-algorithm-that-predicts-what-the-ultra-wealthy-want?utm_term=.jbB9jB2WJ#.mr7kjVQ3P
29. González-Rodríguez, M.R., Díaz-Fernández, M.C., Gómez, C.P.: Facial-expression recognition: an emergent approach to the measurement of tourist satisfaction through emotions. Telematics Inform. **51**, 101404 (2020)
30. https://www.30secondstofly.com/ai-software/ultimate-travel-botlist/#3_Lola_an_iPhone_app_connecting_users_to_travel_agents
31. Gretzel, U., Sigala, M., Xiang, Z., Koo, C.: Smart tourism: foundations and developments. Electron Mark **25**(3), 179–188 (2015)
32. Tsaih, R.H., Hsu, C.C.: Artificial intelligence in smart tourism: a conceptual framework. In: Artificial Intelligence (2018)
33. Tussyadiah, I., Miller, G.: Perceived impacts of artificial intelligence and responses to positive behaviour change intervention. In: Information and Communication Technologies in Tourism 2019, pp. 359–370. Springer, Cham (2019)
34. Impacts of COVID-19 on global tourism industry: a cross-regional comparison
35. Jaipuria, S., Parida, R., Ray, P.: The impact of COVID-19 on tourism sector in India. Tour. Recreat. Res. **46**(2), 245–260 (2021)

36. Chandel, R.S., Kanga, S., Singh, S.K.: Impact of COVID-19 on tourism sector: a case study of Rajasthan, India. Aims Geosci. **7**(2), 224–243 (2021)
37. Prentice, C., Dominique Lopes, S., Wang, X.: The impact of artificial intelligence and employee service quality on customer satisfaction and loyalty. J. Hosp. Market. Manag. **29**(7), 739–756 (2020)
38. Kreiner, N.C., Ram, Y.: National tourism strategies during the Covid-19 pandemic. Ann. Tour. Res. (2020)
39. Kaushal, V., Srivastava, S.: Hospitality and tourism industry amid COVID-19 pandemic: perspectives on challenges and learnings from India. Int. J. Hosp. Manag. **92**, 102707 (2021)
40. Sharma, G.D., Thomas, A., Paul, J.: Reviving tourism industry post-COVID-19: a resilience-based framework. Tour. Manage. Perspect. **37**, 100786 (2021)
41. Sigala, M.: Tourism and COVID-19: impacts and implications for advancing and resetting industry and research. J. Bus. Res. **117**, 312–321 (2020)

Chapter 7
Study of UAV Management Using Cloud-Based Systems

Sonali Vyas, Sourabh Singh Verma, and Ajay Prasad

Introduction

In recent years, unmanned aerial vehicles (UAVs) have fetched a growing interest in UAVs and their applications, as well as recognition and surveillance, and infrastructure testing [1]. UAVs are becoming increasingly common as the developers and intermediaries have been observing the potential of these winged robots in areas like remote sensing, smart cities, security, disaster management and recovery, surveillance, aircraft testing, border security, etc.

While such low-cost UAVs can be used in many valuable ways, limited processing abilities and low internal storage poses a number of challenges. Actually, UAVs do not fully meet the requirements of computation applications (such as on-board image processing) that include real-time data processing and reliability issues [2]. UAVs are evolving technologies that have the power to transform profitable businesses and the public infrastructures. UAVs accelerate recovery processes after natural adversities and can be used for private transport systems. The upsurge of vigorous UAV systems in cities is because of the invasion of UAV and marketable organizations of multiple UAVs. As the airspace of a UAV aircraft is restrictive, it is necessary to maintain most UAV systems utilizing advanced techniques of collision evading [3].

S. Vyas · A. Prasad
University of Petroleum and Energy Studies, Dehradun, India

S. S. Verma (✉)
Manipal University, Jaipur, India
e-mail: ssverma80@gmail.com

© The Author(s), under exclusive license to Springer Nature Singapore Pte Ltd. 2023
V. Kadyan et al. (eds.), *Deep Learning Technologies for the Sustainable Development Goals*, Advanced Technologies and Societal Change,
https://doi.org/10.1007/978-981-19-5723-9_7

Need and Benefits of UAVs

- **Agricultural Needs**
  Agriculture becomes extremely important as a main source of food production to feed the population in this planet. On the other hand, agriculture provides a lot of benefits to the country such as food and non-food product, transportation, and balancing the environment. The demand for food security creates pressure to the decision maker to ensure our world has sufficient food for the entire world. Thus, the usage of the unmanned aerial vehicle (UAV) is an alternative to manage a farm properly to increase its yield. In order to promote the use of UAV in agriculture to support its sustainability [4]. Agriculture development is important especially for social and economic benefit, where agriculture increases food production, increases net income and improves family productivity, reduces income imbalance between rural agricultural employment and urban income of factory workers, and improves condition of the overall environment [5].

- **Mapping Information**
  The concept of unmanned aerial vehicle (UAV)-based mapping and exposes the current system developments and spatial information industry needs. By exploring the historical concepts and current developments in UAV technology and investigating the use of the technology in various mapping and non-mapping applications and it is found that a lot of advantages and potential technology for the surveying and spatial industry [6]. The role of the surveyor and spatial scientist is challenged to alter traditional practices with new and inventive technology and methods such as geographical information system (GIS) with UAV technology. Typically, it was found that the aviation industry has embraced UAV technology more than the spatial industry in a commercial sense. The actual case is UAV may use digital camera to capture the aerial photographs for productively producing a digital map.

- **Forest Surveillance**
  Considering the forest surveillance at low cost and efficient data acquisition unmanned aerial vehicle's (UAV's) and it can be used for observing critical and important area for intruder activities or to know the current state of any object of interest at forest level [7].

- **Collision Avoidance System**
  Collision avoidance is a key factor in enabling the integration of unmanned aerial vehicle into real life use, whether it is in military or civil application. For a long time, there have been a large number of works to address this problem. Because of their low cost, safety benefit, and mobility, UAVs can potentially replace manned aerial vehicles in many tasks as well as perform well in curriculums that tradition manned aerial vehicles do not. However, as there is no human control, UAVs usage encounters several challenges that need to be overcome; and one of those challenges is collision avoidance. In order to be used, an UAV needs the ability to surely avoid collision with both static and moving obstacles. Whereas UAV collision avoidance shares some similarities with that of air traffics and mobile

robots fundamentally, UAVs surely possess many unique characteristics that need to be considered, making them an interesting research ground [8].

- **Cloud-Based Systems**

  The application of cloud-based systems with unmanned aerial vehicles (UAVs) has attracted great interest for disaster sensing. However, the limited computational capability and low energy resource of UAVs present a significant challenge to real-time data processing, networking, and policy making, which are of vital importance to many disaster related applications such as oil-spill detection and flooding [9]. The data captured for analysis is very much of large amount and which brings challenge to the intermittent and limited network resources. Cloud-based frameworks support the management of for this vast size of data which integrates video acquisition, data scheduling, data offloading, and processing for a stable network state to deliver the scalable and efficient systems. At client-side set of components hosted on the UAV which selectively offloads the captured data to a cloud-based server.

Architecture of Cloud Systems

The UAV's and cloud systems are interconnected to form an efficient sustainable environment for providing the collection of all entities to build a smart city in a whole, one inspiring area like smart cities emergency management which often emerge unexpectedly, will need real-time optimization, as well as the deploying and integrating provisional and flexible fleet of smart UAVs. There is also a requirement of a plan to improve cloud-based data collection and dissemination of information between individuals and ground-support control centers. UAVs are also utilized in preventive surveillance. So, it is necessary that the UAV can be trained to set a limit on how it interferes. This is achieved by reducing its level of noise and visibility, as contrasting to noise and visual functions that may occur during fast operation. In a smart city, using multiple UAVs offers the opportunity to blend their integrated capabilities, which can lead to a richer understanding of the environment. Though, merging data to create novel knowledge and enable UAVs to respond appropriately in a coherent manner leading to higher computing methods. The processing potential that can be carried on the drone board will not be sufficient to make consistent decisions. In addition, events listed with UAVs or provided from data analytics will lead to an explosion of high demand real-time computational requirements [10]. The cloud-based systems are considered as a new hybrid model for implementing UAVs more efficiently. The system uses machine learning algorithms (fuzzy logic) [11] cloud model [12] an effective cognitive process characterized by three expected parameters—mathematical expectations, entropy, and super entropy. The standard cloud prototype is mainly used in many fields like quality testing [13]. The autonomous power of UAVs denotes the capability of UAVs for detection, analysis, communication, planning, and decision-making [14] (Fig. 7.1).

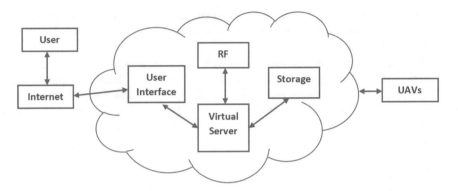

Fig. 7.1 Architecture of cloud-based UAVs [10]

UAV Monitoring and Management System

Smart UAV surveillance systems are significant factor in crowd management as they have verified for effective and profitable resolutions. Utilizing such systems can involve detection of irregularities among the crowd for ensuring public safety, specifically in times of pandemics or social turbulence where technology is intended to replace human factor for ensuring scalability and reducing risk. Conversely, modern framework of autonomous UAV system needs to process embedded data on the edges and resources of the cloud, requiring the transfer and/or re-transfer of embedded data. This data security risk throughout the transfer may jeopardize the technical assistances. So, this involves necessity for an efficient plan to attain a secure framework that takes into account computational capabilities limited to UAV agents and a distributed system. Blockchain, as a distributed network technology, offers a secure, transparent, and efficient network of UAV systems [15].

Blockchain technology is acquainted to accomplish log record monitoring and support the decisions of the transaction monitoring team. The widespread use of UAVs in current years has led to the rapid growth of research facilities related to UAV invention. Also, the development of control systems having ability to perform a wide range of UAV processes is the most explored and dynamic. Presently, researchers are interested in developing control systems that can be called as smart systems that are suitable for solving tasks like planning, prioritization, structural integration, etc. and hence ensuring high levels of autonomy for UAVs. One of the major issues in the design of an intelligent control system involves the various methods and models conventionally utilized in robots and is aimed at resolving tasks like dynamics, signal processing control, location mapping, route planning, etc. [16] (Fig. 7.2).

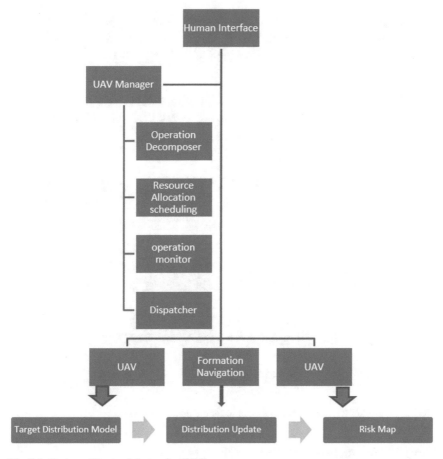

Fig. 7.2 Basic architectural design for UAV's

Self-allocation Distributed Architecture for Collaborative UAVs

The architecture of centralized and distributed control are two types of frameworks of multi-UAV co-operative controls [17]. As well as improving the performance of UAVs, the architecture of shared structures with high reliability levels, low computation, and communication has become the focus of research [18, 19] (Fig. 7.3).

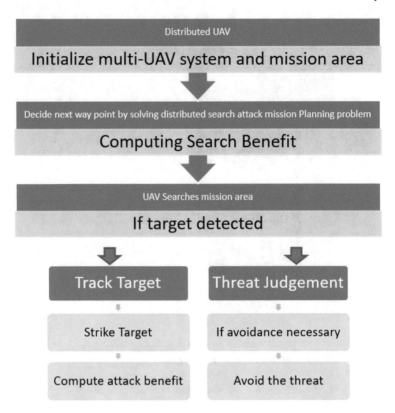

Fig. 7.3 Self-allocation distributed architecture for collaborative UAVs

Cloud Computing and Smart Objects

Recently, UAVs have received substantial attention because of ever-expanding application ranges. Though, building UAV combined systems utilizing conventional
methods in firmly integrated projects which entails the provision of significant development effort, time, and cost. The UAV-cloud infrastructure offers group of services
and APIs in addition to which programs can be built. Because of limited capacity of
these UAVs, lightweight framework is utilized. UAV resources are demonstrated in
resource oriented architecture [20]. Smart objects embedded with these architectures
are well equipped with the configurations on which system can run efficiently. The
development of these kind of smart objects relies the capabilities of these lightweight
UAV's.

UAV-Cloud Framework Layers

The framework of UAV in the cloud computing paradigm enables the development of distributed services of UAVs. Cloud computing is extended for embracing potential computer servers and embedded systems. IoT can be understood as integration of smart objects on the Internet, although offering its resources and services is known as Web-of-Things. UAVs having low power and minimal resources like battery, data processor, and storage devices, can use cloud resources to improve performance in distributed UAVs. Consequently, UAVs do not need to be prepared with robust abilities and have varied resources, therefore implementing this technique with common communication protocols minimizes time and cost of developing applications. UAVs can utilize powerful cloud services, whereas applications of cloud systems may utilize UAVs as service provider. The deployment of services for UAVs is location and capacity dependent.

Therefore, a platform is offered by cloud computing for managing planning and trading services, although UAVs provide special services for specific functions like identifying and putting in action. Such task partition for every object restricts required efforts for building applications beyond such platform. Furthermore, this permits installing multiple UAVs as a plug-and-play system.

Cloud computing and UAV framework:

1. **UAV IaaS**: UAVs and other features are incorporated by this layer. The properties of UAVs comprise of its processors, memory, sensor, and other devices. Additional elements are objects which offer resources like cloud-connected devices and equipment like servers for storage and efficient processors.
2. **UAV PaaS**: It separates infrastructure layer from the application layer. This provides resources as services in the application layer. It permits the integration of cloud services and UAV services to use potential UAV systems. The platform contains UAV services and cloud resources like collaborative equipment planning and resource planning services. The development of collaborative UAVs means the creation of 23 major decision-making skills: equipment planning, job allocation, and systematic implementation [21, 22].
3. **UAV SaaS**: SaaS is a trivial application that is accessible over Internet and assembled at topmost of PaaS via APIs. Designers develop user apps for requesting specific UAV tasks, like software to send request to UAVs for spraying seeds in a particular agriculture zone. The person uses the app to stipulate the location of the agricultural land and make request for spraying of seeds by UAV. It also provides a controlling user interface for keeping track of task advancement and accomplishment. UAVs utilize modified services to detect temperature, take pictures, update, and request other services that require real-time data. These systems can easily be developed over PaaS in similar UAVs due to the division of entity responsibilities (Fig. 7.4).

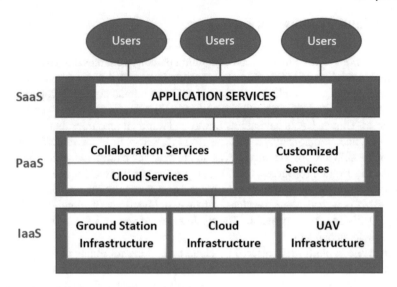

Fig. 7.4 UAV-cloud framework layers

UAV-Cloud Users

The UAV-cloud users utilize cloud computing via APIs and identification based on the rights granted to each user.

1. **End-User**
   End-users are users of SaaS application who launch the UAV functionality. The end-user makes use of the online application software via a browser for requesting a mission that provides certain service limits. Feedback and results are made available on user-friendly interface with some communication abilities.

2. **Application Engineers**
   Engineers or developers register them to the platform for accessing services and APIs for building new applications. They utilize framework applications and facilities to combine them with APIs having pre-defined formats and collaboration agreements to build applications. So, the developer outlines the equipment and UAV services essential to initiate that service. Likewise, the engineers describe the constraints which the end-user must stipulate for requesting a task.

3. **Benefactors of UAV**
   They are UAV holders who are registered for reading and using by engineer to perform specific tasks. As long as UAVs outline the applications based on common interface, and utilize the API for compressing access their data. Registered UAVs are made part of the UAV IaaS cloud and platform APIs.

4. **Managers/Executives**

   Managers own the platform and tracks end-users and resources. Managers function and sustain UAV and cloud processes. Also, utilize techniques and applications for maintaining and controlling the task.

UAV-Cloud Elements

Main emphasis of the study is the layer of "UAV-cloud platform", collaborating "UAVs" into the "cloud" and offering a suitable base for application designing. In conventional growth, the design of applications for explicit systems and it typically means using all the necessary components desired by the system. Such methods are less efficient and also consumes more time and energy. Though, the elements are designed in form of services and combined into applications whenever required. Service comprises of the partnership services essential for collective UAVs provided and implemented in accordance with UAV competences. Creating applications over such services decreases cost and time to improve the integration of UAV systems.

UAV-Cloud Platform Architecture

Cloud platform architecture comprises of two basic architectural models:

- **Web-Service Architecture**: Mainly two kinds of web-service frameworks. Firstly, the design of the standard WS * web-service framework, customer puts request, and objects of response service is processed by the use of "SOAP" and transferred on web by "XML". Secondly, creation of a "Representational State Transfer (RESTful)" web-service framework, which recognizes components using "Uniform Resource Identifiers (URIs)" and "Hypertext Transfer Protocol (HTTP)".
- **Resource-Based Framework for "UAV-Cloud"**: The service-oriented architecture where a benefactor accomplishes the service. Though, the communication provided is services as well as facts like sensory data or home storage data in case of UAVs. Such components are termed as "resources". Thus, for UAV services, SOA is not adequate, whereas more suitable is ROA for representing it.

"REST Architecture"

- Resource identification means, URI to recognize each UAV resources.
- Uniform interface having resources accessible for communication by well-recognized communication semantic (or HTTP), which possess process groups for optimizing resource connections.

- "Self-defining message" accompanied by the HTTP communications, message sets are exchanged among client and server in the verified setup. For machine-focused services, two HTTP-braced media types are there; "XML and JSON". The JSON-based format has received extensive support for embedded programs as a result of its accessibility by users and system and is simple and can be separated directly from JavaScript unlike XML.
- State-less communications, which means no preceding contact data is possessed by the server that impacts any type of the requirements. Thus, every application comprises of necessary data for satisfying requirements. The request details confined in HTTP with an understandable message for the JSON item.

  1. **"RESTful HTTP" Elements**
     An HTTP interface retrieves the resource. The interface includes subsequent specific fragments: (1) processes, (2) content-negotiation, and (3) status codes.
  2. **"RESTful Models"**
     Various model set-ups are available for instantaneous accessible embedded systems resources [23, 24].

- **Building the UAV Layer**: The main concern of converging UAVs with cloud is connection. Many UAVs provision Wi-Fi connections, so they can link with the cyberspace. Furthermore, technology of 3G/4G supports mobiles as well as embedded devices by means of exterior protections making connections with the embedded technology. By the help of this, the UAV receives an exclusive IP address, in order to have a different Internet identity. Like in smart cities, connectivity assumptions are allowed in many application areas. Though, monitoring connectivity of satellite releases up broad application areas. "Internet of Things" helps in studying smart connections and gives them addresses and installs "IPv4 or IPv6" on it. Converging cloud and UAV means that UAV resources are accessible online in a pervasive way to the user. A user is a person utilizing web-based applications, a UAV retrieving third party resources, other systems that works with UAVs or other embedded devices using similar protocol. Consequently, the essential phase is to recognize the services which should be available to customers.
- **UAV Database**: The database contains multiple tables which includes records. Such data is fed during the registration of UAV to the agent. After that a resource table is required for storing the UAV resources. This comprises of the API data of resource that can be considered as resource URI, process, and the benefactor. Such elementary information components are essential for a UAV during its broker registration. Afterward, all the task-assigned UAVs are stored in a separate database.
- **Developing the Layer of Broker**: The deliberation for converging UAVs with cloud is the UAVs' classification and scalability of offering services and resources of API to numerous users. Even though "ROA" is for client–server framework, "RESTful" application provisions services that are integrated, usable, and flexible. In addition, when an engineer develops utility or user creates an assignment, which are attentive to resources of UAV. Thus, RESTful assets develop interactions between the essential services utilizing the agent/broker framework for

taking accountability for UAV allocation for the application. Broker/agent acts as a middle-agent which accepts ads by benefactors based on the abilities as well as service delivery. Afterward, the user requests the seller for a service that specifies the required parameters of a service. Agent then sets comparisons of the demanded service to all accessible ads and regulates provider with perfect match. After that, the benefactor is communicated by the agent and service is requested. When a benefactor receives service request, it renders and returns service as a result to the vendor. Lastly, the result is returned to the client by the broker [25].

- **Front-End Application**: The front-end application is an online software on the client-side. The user utilizes this to enable UAV deployments. It is developed above platform of UAV-Cloud system same as web-application advancement. Afterward, it is executed on cloud system and combined by "collaborative service layer". The request shows the intelligible interface in the web-browser. This visual platform gives the capability to set up task by user, also provides UAV-control. Because of the RESTful freely integrated structure, the layer of applications is developed over service's platform by making use of the developer APIs. Thus, varied utilizations are created by similar UAVs sets operated via broker layer.

UAV Resources Implementation

Implementation comprises of developing resources of UAV and offering APIs. Subsequently, the agent is upgraded for detaching the applicant from the UAV resource. Vendor is linked to a UAV storage-base with resource details. Its resources are used by four UAVs. Each has its own unique RESTful APIs, IP address, and resources. Though, UAVs share same source, describing APIs in similar trend. Intended for convenience, development is achieved by GET method only. Data tables are used in PostgreSQL. The database is accessible to the vendor for accessing, writing, and modifying data by its configuration.

Evaluation of Response Time of UAVs

As a whole, the response time for placing a request for a UAV service differs directly between 180 ms and 470 with an average of 266 ms. By calculating the overcrowding of broker layer, response time of resource was compared for both direct and indirect access. Different response time depends on the resource activity. Thereafter, vendor layer requests resources of UAV by mentioning service name to broker, ensuring the accessibility of the resource requested and to request the UAV as per its corresponding interface and return the results. The average is made ten times per service for each UAV.

UAV-Cloud Versus Other Related Solution

There are many benefits to easily develop UAV applications, severing UAV service responsibilities and integrating them. The UAV-cloud platform overcame conventional peer-to-peer RF connectivity which has recognized many performance and development barriers. Furthermore, creating a unique UAV system utilizing conventional methods takes effort as well as time as it needs each UAV knowledge and planning. The UAV performance is also restricted to certain machines. Additionally, in the case of an "RF" connection, the location of user must be in the equipment range. It is considered that UAVs are part of the web servers to benefit from ubiquitous cloud computing and to facilitate the utilization of web tools and processes to improve UAV interoperability systems. Subsequent cloud-web advancement triggers the ability to build desktop-related applications as well as mobile UAV applications. Moreover, such applications are accessible outside of operating system. UAVs provide not only services as well as resources. "RESTful ROA" implementation is a simple, renewable, and freely integrated web-service. This is appropriate for limited resources of UAV in comparison with composite web-services. Consequently, planning of UAVs is done utilizing "RESTful web-services" for providing services by HTTP interface connectors. Such HTTP APIs are provided by UAVs with their components which are accessible and demanded by the broker layer.

Whereas, execution of UAVs has some drawbacks. It did not specify the broker's rating and number of UAVs dealing with it. It comprises of a large number of applications that can be processed at the same time. Additionally, the influence of similar requests during response time and the way broker behaved is not investigated. The implementation of model used fixed devices; so, the flexibility and spatial factor was not used. Only the pull model was executed. A push-up subscription model was thought to be available. Although the API provisions different devices, UAV implementation is based on the same Arduino devices with varied resources.

UAV cloud, which is private, offers services and structure through its society only and may be handled by the society or third party. The society manages the UAVs and plans can be made as per their needs and performance. In contrast, UAV which are public, uses an open network for offering services. This also opens platform for public business applications of UAV, in which the user is not in charge of UAV management instead gets benefits from their use. Though some documents refer to only a segment of this feature and no standard platform is planned for UAV components utilizing power of cloud computing.

Conclusion

A framework for deploying distributed UAVs is proposed. It offers services to facilitate app expansion. Like this, the integration of UAVs and cloud computing model to offer universal admittance for their resources and services. In "resource oriented architecture" resources and services of UAVs are pursued, this is a novel pattern for scalable web-service creation and freely integrated communication amid services. Therefore, they are accessible as "representational state transfer" services by means of HTTP. In addition, study suggests utilizing vendor properties for increasing efficiency by separating tasks. Consequently, it splits the ideas and performance of the applicant from the providers. It also accepts the responsibility of assigning the application to the existing and appropriate UAVs. For analyzing the discussed structure, equipment of UAV is developed as a sub-payment system and then offered them with Internet connectivity. Subsequently, the same service identifiers and interfaces were established with RESTful application programming interfaces (APIs). A broker service and data containing details of enumerated UAVs and equipment are set. The system elements are verified utilizing the browser interface by allocating a request for help from the vendor to locate and request the service in the appropriate UAV. Tests were performed to obtain data from UAVs and to request actions from them.

Future Work

UAVs are not the only independent programs. They often interact and interchange information with other programs. The discussed infrastructure could be extended to unlock application capabilities to combine UAVs as well as global systems using the RESTful law. Consequently, app incorporates many components for maximizing the performance and competences. Furthermore, UAV provides huge data sets collected, which enables the "big data technology" to evaluate data for making decisions in diverse applications. Though "RESTful" reconstruction is approved by UAVs, there is still lack in standards. For instance, it does not have the standard format set to represent UAV data and service facts.

References

1. Bhaskaranand, M., Gibson, J.D.: Low-complexity video encoding for UAV reconnaissance and surveillance. In: Military Communications Conference (MILCOM), pp. 1633–1638. IEEE (2011)
2. Koubâa, A., Qureshi, B., Sriti, M.F., Javed, Y., Tovar, E.: A service-oriented cloud-based management system for the Internet-of-Drones. In: IEEE International Conference on Autonomous Robot Systems and Competitions (ICARSC), pp. 329–335. IEEE (2017)
3. Itkin, M., Kim, M., Park, Y.: Development of cloud-based UAV monitoring and management system. Sensors **16**(11), 1913 (2016)
4. Norasma, C.Y.N., Fadzilah, M.A., Roslin, N.A., Zanariah, Z.W.N., Tarmidi, Z., Candra, F.S.: Unmanned aerial vehicle applications in agriculture. In: IOP Conference Series: Materials Science and Engineering, vol. 506, no. 1, p. 012063. IOP Publishing (2019)
5. Othman, K.: Integrated farming system and multifunctionality of agriculture in Malaysia. In: XV International Symposium on Horticultural Economics and Management, pp. 291–296. ISHS Acta Horticulturae 655, Malaysia (2004)

6. Samad, A.M., Kamarulzaman, N., Hamdani, M.A., Mastor, T.A., Hashim, K.A.: The potential of unmanned aerial vehicle (UAV) for civilian and mapping application. In: IEEE 3rd International Conference on System Engineering and Technology, pp. 313–318. IEEE (2013)
7. Saadat, N., Sharif, M.M.M.: Application framework for forest surveillance and data acquisition using unmanned aerial vehicle system. In: International Conference on Engineering Technology and Technopreneurship (ICE2T), pp. 1–6. IEEE (2017)
8. Pham, H., Smolka, S.A., Stoller, S.D., Phan, D., Yang, J.: A survey on unmanned aerial vehicle collision avoidance systems (2015). arXiv:1508.07723
9. Luo, C., Nightingale, J., Asemota, E., Grecos, C.: A UAV-cloud system for disaster sensing applications. In: IEEE 81st Vehicular Technology Conference (VTC Spring), pp. 1–5. IEEE (2015)
10. Cochez, M., Periaux, J., Terziyan, V., Kamlyk, K., Tuovinen, T.: Evolutionary cloud for cooperative UAV coordination. In: Reports of the Department of Mathematical Information Technology. Series C, Software engineering and computational intelligence, (1/2014) (2014)
11. Feng, Y., Liu, S., Xie, W.: Autonomous capability evaluation of ground-attack UAV Based on cloud model and combined weight theory. Math. Probl. Eng. **2021** (2021)
12. Gong, X., Yu, C.: Improved TODIM approach for alternative evaluation based on cloud model. Electron. Electr. Syst. Eng. **40**(7), 1539–1547 (2018)
13. Wang, D., Liu, D., Ding, H., et al.: A cloud model-based approach for water quality assessment. Environ. Res. **148**, 24–35 (2016)
14. Liu, S., Ru, L., Wang, K.: New progress in autonomous evaluation methods of UAV. Aero. Mis. Jou **2**, 43–49 (2018)
15. Xiao, W., Li, M., Alzahrani, B., Alotaibi, R., Barnawi, A., Ai, Q.: A blockchain-based secure crowd monitoring system using UAV swarm. IEEE Network **35**(1), 108–115 (2021)
16. Emel'yanov, S., Makarov, D., Panov, A.I., Yakovlev, K.: Multilayer cognitive architecture for UAV control. Cogn. Syst. Res. **39**, 58–72 (2016)
17. Cummings, M.L.: Operator interaction with centralized versus decentralized UAV architectures. In: Handbook Unmanned Aerial Vehicles pp. 977–992 (2015)
18. Cristian, R.A., David, C.: Constrained multi-objective optimization for multi-UAV planning. J. Ambient Intell. Hum. Comput. **10**(6), 2467–2484 (2019)
19. Luo, F., Jiang, C., Du, J., Yuan, J., Ren, Y., Yu, S., et al.: A distributed gateway selection algorithm for UAV networks. IEEE Trans. Emerging Top. Comput. **3**(1), 22–33 (2017)
20. Mahmoud, S.Y.M., Mohamed, N.: Toward a cloud platform for UAV resources and services. In: IEEE Fourth Symposium on Network Cloud Computing and Applications (NCCA), pp. 23–30. IEEE (2015)
21. Lemaire, T., Alami, R., Lacroix, S.: A distributed tasks allocation scheme in multi-UAV context. In: IEEE International Conference on Robotics and Automation, 2004. Proceedings. ICRA'04, vol. 4, pp. 3622–3627 (2004)
22. Gerkey, B.P., Mataric, M.J.: A framework for studying multi-robot task allocation (2003)
23. Duquennoy, S., Grimaud, G., Vandewalle, J.-J.: The web of things: interconnecting devices with high usability and performance. In: International Conference on Embedded Software and Systems, 2009. ICESS'09, pp. 323–330 (2009)
24. Bozdag, E., Mesbah, A., Van Deursen, A.: A comparison of push and pull techniques for AJAX. In: 9th IEEE International Workshop on Web Site Evolution, 2007. WSE 2007, pp. 15–22 (2007)
25. Fasli, M.: Agent Technology for E-Commerce, vol. 86. Wiley, Chichester (2007)

Chapter 8
Contemporary Role of Blockchain in Industry 4.0

Shaurya Gupta, Sonali Vyas, and Vinod Kumar Shukla

Introduction

Nowadays technological expertise like cloud computing, fog computing, Internet of Things (IoT), Internet of Services (IoS), cyber-physical systems (CPS) have opened novel paths for commerce besides innovative commercial models. Apart from this, these sorts of technologies offer numerous compensations in terms of enhancement in the field of computerization, efficiency, charge efficacy, consistency, and excellence, besides suppleness for dissimilar commercial and manufacturing sectors. These probable factors led to the fourth-industrial evolution, i.e. Industry 4.0 [1] that aimed at enhancing dissimilar industrialized subdivisions counting manufacturing engineering also. The foremost norm when considering smart manufacturing within the framework of Industry 4.0 involves empowering connectivity amongst numerous industrialized components, amenities, equipment, dealers besides vendors including manufacturing subsidiary commerce also, forming a valued smart industrial manufacturing network within the whole manufacturing value chain [2]. It will be helping in automating, autotomizing besides optimizing procedures thereby growing suppleness, safety besides efficiency which results in cost optimization and augmented productivity. Smart manufacturing network displayed in Fig. 8.1 alters the manufacturing occupational models besides enabling interactions through value chain and amongst its components also. It will be benefiting the manufacturers in developing suppler and improved merchandises. Furthermore, it enables outline in terms of customizable industrialized manufacturing facilities when considering responsive merchandise progress in terms of its development plus introduction of

S. Gupta · S. Vyas (✉)
School of Computer Science, University of Petroleum and Energy Studies, Dehradun, India
e-mail: svyas@ddn.upes.ac.in

V. K. Shukla
Department of Engineering and Architecture, Amity University, Dubai, UAE

© The Author(s), under exclusive license to Springer Nature Singapore Pte Ltd. 2023   111
V. Kadyan et al. (eds.), *Deep Learning Technologies for the Sustainable Development Goals*, Advanced Technologies and Societal Change,
https://doi.org/10.1007/978-981-19-5723-9_8

the product to the market. Though Industry 4.0 skills and technologies proposes numerous compensations when considering industrialized manufacturing segments, a few challenges need addressing in order to attain its maximum profits. The challenges involve connectivity and info interchange amongst dissimilar machineries, components, localities situated across various organizations, units that are a part of manufacturing value chain. A few obstacles include safety, faith, and dependability apart from improved incorporation of the components in value chain [3]. Blockchain initially provided a podium for enabling and supporting usage of digital money like Bitcoin [4]. However, it is an established fact that blockchain is appropriate as well as convenient for varied applications comprising industrialized manufacturing applications also [5] as a chief benefit of blockchain involves enabling a cluster of units to come to consensus regarding definite activities besides registering that agreement also without the help or authorization of any controlling power. Blockchain integrates practices from peer-to-peer system and inculcates the concept of cryptography for supporting a dispersed shared ledger amongst a network of users or officialdoms. The consensus can be in between companies, persons, robotics, smart devices, and software agents in a way that entirely involved approve on its details, and transactions will be carried out very securely and cannot be altered or modified after being attached to the chain. Besides, blockchain permits for generating noticeable irrevocable audit trails, quantifiable constituents, and admittance to comprehensive info concerning transaction dealings logged in the chain, thereby permitting for comprehensive authentications and tracing.

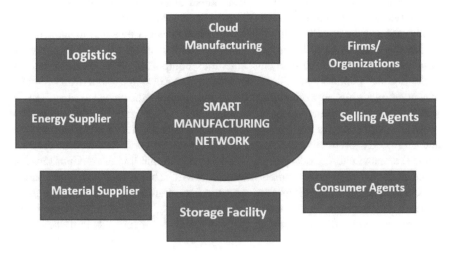

Fig. 8.1 IIoT-enabled smart manufacturing system

Literature Review

In these times of technological advances, the world is getting closer and closer [6] because of rapid developments in the web technology field. Web technologies came into existence in the 70s by the subdivision of Defence Advanced Research Project Agency with the determinations of devolution of dissemination plus message exchange services. The Internet has entirely transformed the existence in terms of education also, as students or researchers are accessing all learning material online or by copying oration on their personal system for learning besides that the content can be sold to educational institutes for purpose of edification. It has made this Author has discussed web development from version 1.0–4.0 [7]. This transformation has evolved web 4.0 comprising artificial intelligence, dealing with sentiments, social technology conjunction which is being showcased in Fig. 8.2 [8].

However, an important application area in area of blockchain involves storing monetary data with the help of safe interchange, apart from varied attempts in exploring apart from extending blockchain applications further than expenditures to additional expanses besides health care, supply chain, industrialized manufacturing, and edification [9]. Blockchain 1.0 is usually allied with cryptocurrency plus expense like Bitcoin. Blockchain 2.0 is allied with computerized digital business by means of smart contracts [10, 11]. Blockchain 3.0 is more fixated on talking regarding wants of the digital society involving smart cities and Industry 4.0 [12]. Industry 4.0 is majorly a vital component of IoT besides other associated technologies. In 2011, German regime was the first to describe Industry 4.0 [13], and it is now

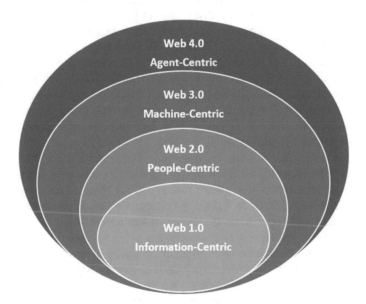

Fig. 8.2 Web evolution from 1.0 to 4.0

extensively acknowledged by commerce and academe both. Industry 4.0 is usually related with a smart workshop or industrial unit kind of setup. The interconnection and digitization of everything and the occurrence of IoT devices in part, facilitate changeover from outmoded industry to smart industry or Industrial IoT. There have been efforts to apply the important concepts of blockchain in IIoT safety [14] apart from that simplifying data assortment [15, 16] besides stowage methods [17, 18]. The authors in [19–21] studied presentations of blockchain for IoT, besides conversed the welfares and confines of applying blockchain in varied applications. Apart from that the reviews which are much concentrated on the application of blockchain in definite commerce were discussed in [22–31] or applications involving edge computing [32] plus disseminated network safety services [3]. In 2011, Germany presented the concept of Industry 4.0 for the year 2020 [33]. Industry 4.0 leads to disseminate extremely computerized besides active manufacturing network systems with technological enablers responsible for the necessary implementation.

Enablers like:

- Industrial Internet of Things
- Blockchain
- Big Data
- Humanoid Device Communication
- Open-Source Software.

Computerization of manufacturing structures is achieved by help of interrelated cyber-physical systems (CPS) in Industry 4.0, which permits the industrialized substructure besides manufacture procedures leading to a transformed independent plus vibrant system [34]. Units, which are a part of incorporated network, communicates apart from acting as smart devices, which unconventionally work and communicate amongst themselves to attain mutual aim [35]. Information and communication technologies (ICT) play an important role while considering sustainability in terms of industrial development in order to support universal financial, communal, in addition to ecological sustainability [36]. In [37] Industry 4.0 discusses some main paradigms such as the first one is about the smart product, which has a control of the capitals besides coordinates the industrialized procedure completely. Second paradigm involves the use of smart machine, i.e. cyber-physical system, where conservative industrialized manufacturing procedures transit itself into disseminated, adjustable, supple, and self-managing production channels. Last paradigm involves an amplified processor, adding to the elasticity plus competence of manual operatives in manufacturing or industrialized organization. It purposes at supplementing employees' competence besides providing supportive work atmosphere, which reshapes the manufacture cycles with the help of human mechanism interfaces permitting association amongst the complete industrial environment. Real-time distribution of info amongst numerous objects or units via a digital network with data stowage competence in terms of large-scale processors remotely is viable with the help of Internet. Industry 4.0 signifies the subsequent phases regarding development of outdated workshops to definite smart workshops, which are more effective while considering resource administration and being quite supple in adaption of changing

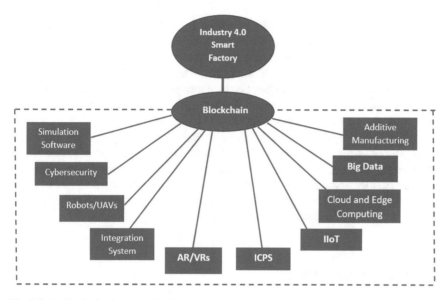

Fig. 8.3 Technologies benefitted by Industry 4.0

manufacturing necessities [38]. The perceptions behindhand Industry 4.0 were first defined by German government in the year 2011 [39, 40]. However, the implemented concept of Industry 4.0 is very much alike to Industrial Internet of Things (IIoT) [41], Internet Plus [42], or made in China 2025 [43]. Industry 4.0 includes group of data from various segments of value chain. Industry 4.0 standard recommends usage of technologies, which enables independent communications amongst manifold manufacturing device units disseminated all over a workshop with the help of Internet. The examples of empowering technologies which is being showcased in Fig. 8.3.

Industry 4.0 prototype nurtures the application associated to varied technologies for enabling the evolution of industrial unit existing design to designs wherein all units are involved in industrial procedures interchange information, i.e. Peer-to-Peer (P2P) network. The most promising expertise, which is applicable in industrialized environment, is blockchain, which was generated from the concept of cryptocurrency. Bitcoin [44] for the purpose of creating decentralized applications, which are very helpful in tracking and storing transactions, which are carried out by numerous users plus devices.

IIoT

IIoT involves the combination of machineries, processors, and persons empowering smart industrialized processes, by means of progressive data analytics for transformational commercial outcomes [45] as it empowers the incorporation of WSN, communication protocols besides Internet infrastructure by empowering smart industrial operations for observing, exploration, and administration. The inter-network of components in manufacture system framework aids in computerization of industrialized manufacture besides increases intellect, effectiveness, and security [46]. IIoT architecture has three layers, the first one being the physical layer consisting of physical device units like sensors, actuators, and industrialized apparatus. Communication layer involves the use of actuators in addition to wireless sensor networks (WSNs), 5G, and machine-to-machine (M2M) message exchange, which facilitates incorporation of varied hardware device units in physical layer for the purpose of industrialized engineering and computerization. These layers construct a CPS for supporting industrialized besides manufacture application layer when considering progress of smart factories, etc. Networking and control frameworks of the cyber-systems empowers the smart process of the manufacture systems.

Manufacturing Industry

In it, constructors have to issue procedural handbooks for the products that needs to be disseminated in the preservation subdivisions. Technical accounts need to be unconfined apart from being updated in a timely manner, which is a tiresome process involving a great amount of official procedure. By implementation of blockchain technology, the technical publication utilizes the framework, which is accessible to blockchain users. Blockchain has suggestively improved the operational competence in industrialized commerce by means of data upload on shared ledger. Considering the case of auto business, tracing of spare parts is quite important, as the obtainability of those parts when considering a real-time scenario is never known. Blockchain aids in resolving the matter by apprising the pertinent info of spare parts on shared ledger and it is accessible to every entity involved like car constructer, warehouse suppliers, etc., as presented in Fig. 8.4.

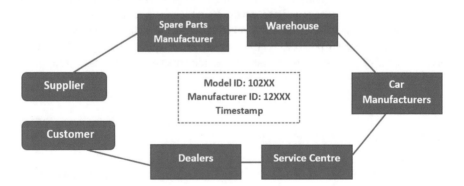

Fig. 8.4 Standby parts locating using blockchain in motor Industry

Blockchain Technology and Industry 4.0

Industry 4.0 expertise may be benefitted by the implementation of the blockchain, nevertheless the usage also presently poses a challenge in various facets. Application of blockchain can assist cloud computing solutions providing redundancy of storage requirements, although local blockchain deployments are presently very tough to duplicate on IIoT nodes because of the storage and operational limitations. Thus, when this technology is used in convergence to Industry 4.0 application, it becomes important to deliberate the trade-off amid the usage pros and cons of blockchain.

- Horizontal and Vertical Incorporation Structures
  Horizontal and vertical incorporation is vital for facilitating information exchange within industry and to interconnect with dealers and customers. Conventionally, such incorporations are delivered via Manufacturing-Execution Systems, Product Lifecycle Management, and Organizational Planning; then again need an advanced level of amalgamation with Industry 4.0, so that the above framework can be public. Like, horizontal integration based on blockchain with another industry partners (hence need for added integration, which is often expensive) is given in the electronics industry [47].

  One such chapter proposes the utilization of a blockchain for developing a substituting supply of power in a cooperative way. This type of association takes place between the manufacturer and a team of engineers. Offers of switching designs based on multichain [48] published by producers on Blockchain, technologists' study and evaluate such proposals and decide completion of the proposed reward. The parallel integration essential in industrial manufacturing and supply chain methods aims for blockchain-based improvements, because of distributed transactions and data management competences. In supply chain, there are interchanges which are essential for recording transactions and specify proprietorship. The recording of transaction helps in keeping track of possessions and offers transparency to third party.

  For example, in relation to blockchain-based vertical connectivity, blockchain has been assessed by implementation in terms of supply chain of complex materials manufacturing commerce that constructs merchandises for businesses such as automobile, airways, or else railways [49]. The operation of blockchain to automatically negotiate the resources provided by the 5G infrastructure. Reservation on-demand can be conducted by mobile networks and operators in a dynamic and easy manner.
- Inter-disciplinary Cyber-Physical System (ICPS)
  ICPS or cyber-physical production system (CPPS) has ability to gathering, processing and storing data, and actions to regulate physical procedures. The various elements of ICPS are still substantially dispersed through the smart factories with decentralized data processing and analysis [50] and taking real-time decisions. As a result of the decentralized ICPS and its necessity for data retrieval, blockchain proves to be a good counterpart for such system. Blockchain proves to be backbone for ICPS in varied organizations. Many researchers focused on

growing CPS reliability by generating a reputation system and providing suitable rewards to different organizations collaborating with the system. Furthermore, many scholars proposed to improve production procedures by distributing and then generating an ICPS to manage and regulate them [51]. In this situation, the researchers focused on the benefits in utilization of an ICPS blockchain like coordinating local real-time methods, but the problem still needs to be talked before marketable extension, Transaction Serialization, that generates a blockage of performance and affects scalability.

Benefits of Blockchain Technology in Manufacturing Systems

- Blockchain Computation
  Blockchain is a combination of chain structure, consensus algorithm, cryptography technology, and smart contracts. It contains decentralized efficient data block. Each data block comprises timestamp with a linkage to preceding block. Blockchain covers complete historic information so that every transaction in the database can be identified as a source. Blockchain is addressed as a secure and shareable computing model. Blockchain implementation in manufacturing is generally established on the basis of structure offered by the typical blockchain frameworks. Blockchain offers collection of distributed data structures, collaboration methods, and computation models that provide security measures for exchanging information, services, or products [52].
- Metrics for Employing Blockchain
  Metrics for operating a blockchain in manufacturing systems includes cybersecurity, decision architecture, system performance, and trust improvement. From a cybersecurity viewpoint, blockchain delivers a flexible and efficient way to manage distributed records on the Internet. The chain infrastructure connects the data blocks consecutively and shields the common ledger from being meddled or replicated in a cryptographic manner. The M1 metric offers decentralized decisions for manufacturers with secure product design, proprietorship certification, and component certification and without intermediary control. In addition, blockchain allows data provision (M2), which is traditionally tough to attain, and makes the system clearer [53].

Challenges

In spite of the advantages offered by blockchain techniques, the development and expansion into Industry 4.0 face important challenges which entail additional study:

- Scalability: The frameworks nominated to provision the applications of blockchain Industry 4.0 will need to resist the enormous traffic generated by such applications. The huge amount can be issues with conventional cloud architecture that has advanced structures that often provide fundamental services like in fog computing architectures [54].

- Unanimity in Algorithm Selection: As consensus algorithm is critical to blockchain efficiency, it must be carefully selected. It is significant to notice that this is not essential to utilize a more consensual (in line with) agreement that may include providing all miners the similar weightage while voting, because in few blockchains, the system will be susceptible to Sybil attacks and solo unit will have ability to regulate the entire blockchain. In addition to PoW consensus algorithms such as those utilized by Bitcoin, like Proof-of-Stake, Practical Byzantine Fault Tolerance (PBFT) [55].

- Energy Efficiency and Latency: Issues related to latency and throughput were earlier defined in association to IIoT in addition to CPSs are induced to different applications of blockchain. For energy efficacy, usage of mining and incompetent P2P principles besides composite cryptographical computation processes contribute to energy consumption, though such factors are challenging with the use of battery-powered devices. In terms of blockchain throughput besides latency, as how the consensus algorithm processes and blocks are embedded in the blockchain affect them. Actually, both factors tend to upsurge latency and significantly decrease throughput, for example, conventional data systems, therefore, it becomes tough to generate real-time responses of events.

Conclusion

Industry 4.0 is a standard that changes the way firms work with inclusion of the latest technologies. Technologies like blockchain, which is effectively implemented in cryptocurrency and can improve Industry 4.0 technology by enhancing security, reliability, stability, efficiency, decentralization, and high-degree automation by smart contracts. This chapter provides exhaustive study of the blockchain benefits in Industry 4.0 technology, and also present challenges. Subsequently offering a standardized approach to determining the implementation of the blockchain is a viable possibility considering the case of Industry 4.0 system. Also, appropriate implementations of blockchain in each area of Industry 4.0 is premeditated, besides some of the foremost challenges. Like this, the article offers a study associated to emerging Industry 4.0 applications and will help software developers to analyse the implementation of blockchain for improving the next generation of secure industrial applications.

References

1. Lasi, H., Fettke, P., Kemper, H.G., Feld, T., Hoffmann, M.: Industry 4.0. Bus. Inf. Syst. Eng. **6**(4), 239–242 (2014)
2. Lu, Y.: Industry 4.0: a survey on technologies, applications and open research issues. J. Ind. Inf. Integr. **6**, 1–10 (2017)
3. Zhou, K., Liu, T., Zhou, L.: Industry 4.0: towards future industrial opportunities and challenges. In: 12th International Conference on Fuzzy Systems and Knowledge Discovery (FSKD), pp. 2147–2152. IEEE (2015)
4. Ozisik, A.P., Bissias, G., Levine, B.: Bitcoin Wiki (2017)
5. Al-Jaroodi, J., Mohamed, N.: Industrial applications of blockchain. In: IEEE 9th Annual Computing and Communication Workshop and Conference (CCWC), pp. 0550–0555. IEEE (2019)
6. Benito-Osorio, D., Peris-Ortiz, M., Armengot, C.R., Colino, A.: Web 5.0: the future of emotional competences in higher education. Glob. Bus. Perspect. **1**(3), 274–287 (2013)
7. Nath, K., Dhar, S., Basishtha, S.: Web 1.0 to web 3.0-evolution of the web and its various challenges. In: International Conference on Reliability Optimization and Information Technology (ICROIT), pp. 86–89. IEEE (2014)
8. Dwivedi, Y., Williams, M., Mitra, A., Niranjan, S., Weerakkody, V.: Understanding advances in web technologies: evolution from web 2.0 to web 3.0 (2011)
9. Taylor, P.J., Dargahi, T., Dehghantanha, A., Parizi, R.M., Choo, K.K.R.: A systematic literature review of blockchain cyber security. Digit. Commun. Netw. **6**(2), 147–156 (2020)
10. Parizi, R.M., Dehghantanha, A.: Smart contract programming languages on blockchains: an empirical evaluation of usability and security. In: International Conference on Blockchain, pp. 75–91. Springer, Cham (2018)
11. Puthal, D., Malik, N., Mohanty, S.P., Kougianos, E., Das, G.: Everything you wanted to know about the blockchain: its promise, components, processes, and problems. IEEE Consum. Electron. Mag. **7**(4), 6–14 (2018)
12. Efanov, D., Roschin, P.: The all-pervasiveness of the blockchain technology. Procedia Comput. Sci. **123**, 116–121 (2018)
13. Meyer, U.: The emergence of an envisioned future. Sensemaking in the case of "Industrie 4.0" in Germany. Futures **109**, 130–141 (2019)
14. Miller, D.: Blockchain and the internet of things in the industrial sector. IT Prof. **20**(3), 15–18 (2018)
15. Huang, J., Kong, L., Chen, G., Wu, M.Y., Liu, X., Zeng, P.: Towards secure industrial IoT: blockchain system with credit-based consensus mechanism. IEEE Trans. Industr. Inf. **15**(6), 3680–3689 (2019)
16. Liang, W., Tang, M., Long, J., Peng, X., Xu, J., Li, K.C.: A secure fabric blockchain-based data transmission technique for industrial Internet-of-Things. IEEE Trans. Industr. Inf. **15**(6), 3582–3592 (2019)
17. Zhao, S., Li, S., Yao, Y.: Blockchain enabled industrial Internet of Things technology. IEEE Trans. Comput. Soc. Syst. **6**(6), 1442–1453 (2019)
18. Fernández-Caramés, T.M., Fraga-Lamas, P.: A review on the use of blockchain for the Internet of Things. IEEE Access **6**, 32979–33001 (2018)
19. Ferrag, M.A., Derdour, M., Mukherjee, M., Derhab, A., Maglaras, L., Janicke, H.: Blockchain technologies for the internet of things: research issues and challenges. IEEE Internet Things J. **6**(2), 2188–2204 (2018)
20. Ali, M.S., Vecchio, M., Pincheira, M., Dolui, K., Antonelli, F., Rehmani, M.H.: Applications of blockchains in the Internet of Things: a comprehensive survey. IEEE Commun. Surv. Tutorials **21**(2), 1676–1717 (2018)
21. Dai, H.N., Zheng, Z., Zhang, Y.: Blockchain for Internet of Things: a survey. IEEE Internet Things J. **6**(5), 8076–8094 (2019)
22. Shen, C., Pena-Mora, F.: Blockchain for cities—a systematic literature review. IEEE Access **6**, 76787–76819 (2018)

23. Fraga-Lamas, P., Fernández-Caramés, T.M.: A review on blockchain technologies for an advanced and cyber-resilient automotive industry. IEEE Access **7**, 17578–17598 (2019)
24. Lu, H., Huang, K., Azimi, M., Guo, L.: Blockchain technology in the oil and gas industry: a review of applications, opportunities, challenges, and risks. IEEE Access **7**, 41426–41444 (2019)
25. Xie, J., Tang, H., Huang, T., Yu, F.R., Xie, R., Liu, J., Liu, Y.: A survey of blockchain technology applied to smart cities: research issues and challenges. IEEE Commun. Surv. Tutorials **21**(3), 2794–2830 (2019)
26. Wan, J., Li, J., Imran, M., Li, D.: A blockchain-based solution for enhancing security and privacy in smart factory. IEEE Trans. Industr. Inf. **15**(6), 3652–3660 (2019)
27. Kassab, M.H., DeFranco, J., Malas, T., Laplante, P., Neto, V.V.G.: Exploring research in blockchain for healthcare and a roadmap for the future. IEEE Trans. Emerg. Top. Comput. (2019)
28. Fernandez-Carames, T.M., Fraga-Lamas, P.: A review on the application of blockchain to the next generation of cybersecure Industry 4.0 smart factories. IEEE Access **7**, 45201–45218 (2019)
29. Musleh, A.S., Yao, G., Muyeen, S.M.: Blockchain applications in smart grid–review and frameworks. IEEE Access **7**, 86746–86757 (2019)
30. Abou Jaoude, J., Saade, R.G.: Blockchain applications–usage in different domains. IEEE Access **7**, 45360–45381 (2019)
31. Yang, R., Yu, F.R., Si, P., Yang, Z., Zhang, Y.: Integrated blockchain and edge computing systems: a survey, some research issues and challenges. IEEE Commun. Surv. Tutorials **21**(2), 1508–1532 (2019)
32. Salman, T., Zolanvari, M., Erbad, A., Jain, R., Samaka, M.: Security services using blockchains: a state of the art survey. IEEE Commun. Surv. Tutorials **21**(1), 858–880 (2018)
33. Bortolini, M., Ferrari, E., Gamberi, M., Pilati, F., Faccio, M.: Assembly system design in the Industry 4.0 era: a general framework. IFAC-PapersOnLine **50**(1), 5700–5705 (2017)
34. Shrouf, F., Ordieres, J., Miragliotta, G.: Smart factories in Industry 4.0: a review of the concept and of energy management approached in production based on the Internet of Things paradigm. In: IEEE International Conference on Industrial Engineering and Engineering Management, pp. 697–701. IEEE (2014)
35. Wu, J., Guo, S., Huang, H., Liu, W., Xiang, Y.: Information and communications technologies for sustainable development goals: state-of-the-art, needs and perspectives. IEEE Commun. Surv. Tutorials **20**(3), 2389–2406 (2018)
36. Munera, E., Poza-Lujan, J.L., Posadas-Yagüe, J.L., Simo, J., Blanes, J.F., Albertos, P.: Control kernel in smart factory environments: smart resources integration. In: IEEE International Conference on Cyber Technology in Automation, Control, and Intelligent Systems (CYBER), pp. 2002–2005. IEEE (2015)
37. Weyer, S., Schmitt, M., Ohmer, M., Gorecky, D.: Towards Industry 4.0-standardization as the crucial challenge for highly modular, multi-vendor production systems. Ifac-Papersonline **48**(3), 579–584 (2015)
38. Prifti, L., Knigge, M., Kienegger, H., Krcmar, H.: A Competency Model for" Industrie 4.0" Employees (2017)
39. Hermann, M., Pentek, T., Otto, B.: Design principles for Industry 4.0 scenarios. In: 2016 49th Hawaii International Conference on System Sciences (HICSS), pp. 3928–3937. IEEE (2016)
40. Da Xu, L., He, W., Li, S.: Internet of things in industries: a survey. IEEE Trans. Industr. Inf. **10**(4), 2233–2243 (2014)
41. Wang, Z., Chen, C., Guo, B., Yu, Z., Zhou, X.: Internet plus in China. IT Prof. **18**(3), 5–8 (2016)
42. Glaser, B.S.: Made in China 2025 and the Future of American Industry. Center for Strategic International Studies
43. Nakamoto, S.: Bitcoin: A Peer-to-Peer Electronic Cash System, p. 21260. Decentralized Business Review (2008)
44. Swan, M.: Blockchain: Blueprint for a New Economy. O'Reilly Media, Inc. (2015)

45. Matthyssens, P.: Reconceptualizing value innovation for Industry 4.0 and the Industrial Internet of Things. J. Bus. Ind. Mark. (2019)
46. Gupta, S., Vyas, S., Sharma, K.P.: A survey on security for IoT via machine learning. In: International Conference on Computer Science, Engineering and Applications (ICCSEA), pp. 1–5. IEEE (2020)
47. Yan, Y., Duan, B., Zhong, Y., Qu, X.: Blockchain technology in the internet plus: the collaborative development of power electronic devices. In: IECON 2017–43rd Annual Conference of the IEEE Industrial Electronics Society, pp. 922–927. IEEE (2017)
48. Adam, K.: White Paper: "Project Hurricane"-or How to Implement Blockchain Technology in German Real Estate Transactions (2017)
49. Backman, J., Yrjölä, S., Valtanen, K., Mämmelä, O.: Blockchain network slice broker in 5G: slice leasing in factory of the future use case. In: Internet of Things Business Models, Users, and Networks, pp. 1–8. IEEE (2017)
50. Harrison, R., Vera, D., Ahmad, B.: Engineering methods and tools for cyber–physical automation systems. Proc. IEEE **104**(5), 973–985 (2016)
51. Afanasev, M.Y., Fedosov, Y.V., Krylova, A.A., Shorokhov, S.A.: An application of blockchain and smart contracts for machine-to-machine communications in cyber-physical production systems. In: IEEE Industrial Cyber-Physical Systems (ICPS), pp. 13–19. IEEE (2018)
52. Andrews, C., Broby, D., Paul, G., Whitfield, I.: Utilising financial blockchain technologies in advanced manufacturing (2017)
53. Lee, J.H., Pilkington, M.: How the blockchain revolution will reshape the consumer electronics industry [future directions]. IEEE Consum. Electron. Mag. **6**(3), 19–23 (2017)
54. Preden, J.S., Tammemäe, K., Jantsch, A., Leier, M., Riid, A., Calis, E.: The benefits of self-awareness and attention in fog and mist computing. Computer **48**(7), 37–45 (2015)
55. Castro, M., Liskov, B.: Practical byzantine fault tolerance. In: OSDI, vol. 99, no. 1999, pp. 173–186 (1999)

Chapter 9
SDGs Laid Down by UN 2030 Document

Vishakha Goyal and Mridul Dharwal

Introduction

Development is a complete phenomenon in itself. In the longer run, any economic development can't be sustained without social and environment development. To achieve this broader objective, UNDP had given a vision to its 189 member nations. In the year 2000, total eight Millennium development goals were introduced/framed and given to its member nations to chase. These eight MDGs range from reducing poverty by 50% to providing basic hygiene and education to all. The target was supposed to achieve by the year 2015. The UN was working with all governments to build a base for providing safeguard to all vulnerable groups. With the achievement in Millennium development goals, the United Nations started preparing a sound ground for next step which can provide a wider horizon to its member nations [1]. In September 2015, the General Assembly of United Nations adopted the agenda for Sustainable Development. The target to achieve these goals is 2030. When the world is moving for the economic development, it is necessary to protect environment so that production can be protected. Economic development is incomplete if human development is centered in a particular region or race. Every single person should get his/her share of joy, good health, development and peace. However, it is not the first time when United Nation start working beyond economic development at international level. The concept Human Development Index (HDI) was formulated to create awareness among nations to think beyond per-capita income and GNI. Equal distribution of resources has a deep linkage with sustainable development.

V. Goyal (✉)
Sharda University, Greater Noida, India
e-mail: vishakha.j1@gmail.com

Vivekananda College, Delhi University, New Delhi, India

M. Dharwal
School of Business Studies, Sharda University, Greater Noida, India

© The Author(s), under exclusive license to Springer Nature Singapore Pte Ltd. 2023
V. Kadyan et al. (eds.), *Deep Learning Technologies for the Sustainable Development Goals*, Advanced Technologies and Societal Change,
https://doi.org/10.1007/978-981-19-5723-9_9

HDI ranking was calculated on different parameters like average year of schooling, life expectancy and income [2]. This ranking was also providing awareness to nations that just a concentrated effort to increase GNI is not sufficient. A major drawback in this HDI was, it was issued once a year. To overcome this deficiency and creating a nudge to all nations, a joint initiative of WHO and UNICEF was constituted named the joint monitoring program (JMP). This agency is responsible for releasing the data on different development parameters of its member countries. This joint monitoring program has given a new dimension in this approach. It collects the data from different governments, international agencies and analyzes this data to give a holistic picture of regional development in the world.

Sustainable Development Goals (SDGs)

A new dimension is added to SDG when moving from MDG is equality. The basic principle of SDG is "leaving no one behind", this means there is special emphasis on vulnerable groups like physically challenged, women and other marginalized communities. During MDG, it was observed that the target was achieved at the global level but there were high differences when compared with in the nation. Hence this time, a wider perspective has taken to be achieved in a limited time frame. This global target has to be achieved by the year 2030. There are total 17 major goals with their respective targets within each goal [3].

SDG 1: No Poverty
The major thrust of this goal is "End poverty in all its forms everywhere", which means that attack poverty from all dimensions. There is a goal to bring income of the people above 1.25$ a day [4]. This means that reduction of poverty without any gender discrimination [5]. Target 1.2 is focused on reduce the poor by half including men, women and children. Another dimension of this goal is to provide social security (target 1.3) and equal rights to economic resources (Target 1.4) to all vulnerable communities. While providing the equal opportunities, the access to microfinance, inheritance and property rights can't be neglected. The worst affected people of any natural disaster/environment shocks are poor so it is the responsibility of the nation to provide some cover for all poor to absorb/resist from these problems. Nation is also responsible to create a policy framework to support gender-sensitive development (target 1.5).

SDG 2: Zero Hunger
Food is a very basic requirement for human survival; hence, hunger is the first threat to mankind.

Hunger can be defined when people are facing severe food shortage and they have to remain empty stomach for days. Nature has given enough resources everywhere, still approximately 2.37 billion people remain empty stomach for days due to shortage of food in 2020 [6]. This goal is dedicated to reduce the number of vulnerable. This includes children, infants and women (Target 2.1). Malnourishment among infant and children is the most dangerous for the development of any nation. Stunting and wasting is a common outcome of malnourishment, which refrain children to become healthy adult. Target 2.2 is reducing the number of factor responsible for stunting and wasting. That means, malnourishment should be reduced among pregnant women/lactating mothers, adolescent girls and children less than 5 years of age. The easiest solution to fight from hunger is to increase the production efficiencies in agriculture. Target 2.3 is to increase the family income of small farmers through equal access to land, market which can make the entire area food secure. Climate change is a truth of the day and agriculture will be its first sufferer so target 2.4 is to develop sustainable agriculture practices for food production. This means a complete framework to fight from floods, drought and improving the soil quality. This target can't be achieved without scientific development in the production of seeds, wild species and animals. Another dimension of this goal is development in rural infrastructure which is necessary for smoother trade in agriculture including exports and imports. It is also experienced that hunger is also an outcome of food wastage, development in supply chain can help in reducing food wastage. High perish nature and bulkiness make difficult to transport agricultural goods easily, hence smart supply chain including packaging is required to make entire sector more efficient. This can reduce food prices, food shortage and ultimately malnourishment.

SDG 3: Good Health and Well-Being
Maternal mortality is still a serious concern particularly for developing countries. Goal 3 is focused to reduce the maternal mortality ratio to less than 70 per 100,000 live births. The another dimension of this goal is to reduce neonatal mortality to 12 per 1000 live birth and under-5 mortalities to 25 per 1000 live birth (target 3.2). These are the basic parameters of health, besides there is target to reduce the communicable diseases and non-communicable diseases like AIDS, Malaria, tuberculosis and others (Target 3.3). For the first time, mental health is also given its due importance in global goals, and drug abuse and mental health should be reduced (Target 3.4, 3.5). For making the world a better place to live, it is necessary to focus on family planning methods and universal health coverage including basic vaccination (target 3.6, 3.7, 3.8). Tobacco consumption should be reduced in all countries.

SDG 4: Quality Education
The thrust area of this goal is to provide inclusive and equitable quality education to all which ensure effective learning outcomes. Education is not a new area of focus by United Nations but under SDG inclusive education is targeted which can accommodate all (without gender or economic biasness). This time the focus is not just limited to primary education or the literacy but imparting education to tertiary level which can create equal opportunities for all [7]. Education is a basic medium for human development. It also provides awareness in the society to be sensible toward environment and society. Education will also encourage the concept of non-violence, cultural diversity and global citizenship [8].

SDG 5: Gender Equality

Gender equality is must to make a balanced world. Eliminate gender discrimination in its all form is the major theme of SDG 5. Violence against women and girls can disturb the balance of the society. Safe environment for women (free from sexual/physical and mental abuse) not only can promote economic harmony in the society can also bring mental peace in the families. Forced marriage/genital mutilation is also considered a form of sexual abuse. A nation should provide strong public policies to reduce such kind of social norms. This target can be achieved with equal participation of women in leadership and economic opportunities. Clear laws should be formed to provide sexual freedom to women above 15 years [3]. The decision of sexual reproduction should also remain with women between the age group of 15–49 years. Women should be given equal property rights and inheritance. Technology, particularly communications technology, can promote women empowerment. Hence, use of mobile should be increased among women.

SDG 6: Clean Water and Sanitation

The first requirement for human survival is drinking water and facilities to defecate in safe manner. Goals 6 of SDG address this basic issue. By 2030, it is necessary to provide these two basic facilities to its entire citizen without any discrimination. Safe and affordable drinking water is becoming a challenge for many regions. Climate change has added difficulties in it. Similar is about the safe option to defecate. To end open defecation is a key area while framing SDG, the horizon to define safe sanitation has increased [9]. The concept starts from giving the access of toilets to the all without any discrimination. No one should be left for this basic right that means every gender and every caste/social group/physically challenged should get equal access to toilets. The gender issue has been specifically focused as toilets also provide safety to women, young girls, physically disable and transgender against different social crimes. Access can't solve the entire problem so the safe sanitation should ensure that, the human waste should not remain in direct touch of human through any mean. If this waste moves through open drains or dumped in solitaire place, it is equally harmful to the human and environment health. Hence under SDG, the due emphasis has given on development of a system for treatment of sludge.[1] The third target under SDG 6.2 is on hand wash. Most of the virus/bacteria enter in the stomach through hands or nails. To reduce the risk of infection, it is important to wash hand after defecation and before consuming meals with water and some disinfectant like soap.[2] Safe sanitation has become even more important to women and girls equitable sanitation facilities, including menstrual hygiene should be provided to all females. Throwing sludge into rivers, pond and other water bodies contaminated drinking water and hence become a reason of many diseases so sludge management has taken care in safely managed sanitation. Wastewater should be treated safely before mixing with natural water bodies. Integrated water management is also taken care to ensure water availability in the coming years.

---

[1] https://www.who.int/news-room/fact-sheets/detail/sanitation.

[2] JMP methodology: 2017 update and SDG baselines.

SDG 7: Affordable and Clean Energy
Energy is a basic resource for the development of the society. This goal is dedicated to provide universal access to affordable energy to all. Use of clean fuel should be encouraged through high share of renewable energy. It is targeted to increase the proportion of renewable energy to double the global rate of improvement in energy. International treaties can facilitate the production of renewable energy and increase energy efficiency. With this technology advancement, the world can reduce global emissions and dependency on fossil fuel.

SDG 8: Decent Work and Economic Growth
Economic growth is a long-term objective for every nation. There is target to achieve minimum of 7% growth in gross domestic product per annum for low and middle-income countries. To achieve this objective, technological upgradation and innovation can increase production efficiencies to provide high value addition. Particularly for least developed countries, unemployment is a major problem. By increasing production opportunities, job opportunities could be increased. There is need to develop strong public policies to motivate entrepreneurship activities, hence financial services should be focused (Target 8.3). Economic growth should not accompany with environment protection (target 8.4). Further this goal is targeting about equal wages for men, women and persons with disabilities (target 8.6). For the economic development, child labor or human trafficking should be addressed in planned way. Labor laws should be framed in compliance of International Labor Organization (ILO) to help the vulnerable (target 8.8). Job creation at local level can also promote tourism, which would be helpful to increase job opportunities at regional level (8.9).

SDG 9: Industry, Innovation and Infrastructure
To achieve the long-term goal of an economy, sound infrastructure has to be provided to support economic and human well-being. In modern times, international trade plays a vital role; hence, infrastructure should be providing both at regional and international level (9.1). Mass employment opportunities can be created through manufacturing sector; hence, this sector should be given more importance. Better infrastructure can increase the resource efficiency with low emissions. To achieve this target, it is necessary to focus on research which can provide resilient infrastructure in developing countries.

SDG 10: Reduce Inequality
An inequality in any form creates barriers in the enhancing natural capabilities of an individual. Hence, reducing inequalities is soul of sustainable development goals. Inequalities have many dimensions and direction. It could be within a country or among different countries. Income inequality is an important dimension of all inequalities. Providing equal opportunities of trade can help in reducing relative income inequalities to all countries. Wealth inequalities can be targeted by providing income growth (higher than the country average) to bottom of the pyramid (bottom 40% of the population). This will increase the social, economic and political inclusion of poor. A country can't grow in its true sense if the fruit of growth will not reach

to the bottom of pyramid. Beside income and wealth, there are other inequalities like sex, physical/mental disability, age, caste or race. Under SDG 10, a country should ensure to protect the rights of these vulnerable groups. For this, strong public policies and laws are required to provide equal opportunities in wages or social protection. For greater global equalities, developing countries should be given more representation on different international forum. The recent COVID-19 pandemic has highlighted the vulnerability of migratory labor. Hence, a country is required to facilitate the international and regional migration of workforce with planned. By 2030, reduction in transaction cost of remittances to less than 3% is another agenda in this.

SDG 11: Sustainable Cities and Communities
This goal has total 7 targets. The target is to ensure access of adequate, safe and affordable housing to live. Slums are very common in emerging developing cities. This target pledge to eliminate slums and instead every nation ensure to provide basic affordable housing facility to all. These affordable houses might be little far from the densely populated cities so a well-developed transportation system can reduce this gap. Hence by 2030, it is important to provide affordable and safe public transportation system to provide a successful framework to cater the special need of women, children, persons with disabilities and older persons [10]. For culturally rich countries like India, it is also important to protect natural heritage. In this way, the ill effects of climate change and man-made disaster can be reduced. The next target under SDG 11 is to develop people from natural calamities including water-related diseases. By 2030, it is important to reduce per-capita emissions to safeguard the cities from negative environmental impacts. For this, municipalities have to provide special focus on maintaining air quality through proper waste management. To achieve this broader goal, three small targets have also included.

SDG 12: Responsible Consumption and Production
The psychological law of consumption says that with increase in income, the consumption will also increase. The natural resources are limited in the world and high-income countries with higher income will consume more. Hence this goal is dedicated to balance the nature of consumption in such a way that everyone will get its due share of goods and services. This goal has eight target and three sub-targets. To achieve sustainable management of resources so that natural resources can be used in efficient manner. The immediate crisis is in food products. There are some people who are facing hunger and starvation while other segment has enough food to waste. It is one target under this goal to reduce per-capita global food waste by half by 2030 [11]. The nature of agricultural food is highly perishable so if supply chain in agricultural food crops is well developed, it can reduce the wastage of food. Excessive chemicals in pesticides or manure can provide adverse impact on human health. Hence, it is also important to reduce maintain some standards agreed world-wide. This is also help to reduce emissions and provide safeguard to air, water, soil and the environment. The 3 'R', i.e., reduction, recycling and reuse can decrease the waste so that natural resources can be utilized to produce for poor's. A well-defined public procurement policy can also increase the sustainability which is as per the

national priorities. This all can be achieved only if people are aware about sustainable lifestyle to increase the harmony with nature. To promote sustainable lifestyle and sustainable tourism, it is important to create jobs locally and promotes products and local culture.

SDG 13: Climate Action

As per the IPCC Special Report (2018), there will be increase in the global temperature by 1.5 °C from pre-industrial levels due to greenhouse gas emission at global level [12]. To maintain the industrial production and reduce the poverty, it is important to reduce the emissions of GHCs. It is important to form strong public policies to monitor the emissions to reduce the impact of climate change. Awareness among people and improved technology in production can work as catalyst. Developed countries can invest in improved technology and replace in their industries, hence it is the responsibility of developed countries to help the developing countries in this. United Nations Framework Convention on Climate Change, it is important to create a fund of $100 billion annually by 2020 from all sources to address the needs of developing countries for necessary actions to be taken by all.

SDG 14: Life Below Water

The marine life has high economic, industrial and environmental value. The coastal resources is providing employment to many. This industry constitutes to 5% of global GDP (UN, Sustainable development goals, Goal 14 webpage http://www.un.org/sustainabledevelopment/oceans/). Pollution created by humans choking the marine life, around 40% of the oceans has been suffering from high pollution which becomes a threat to marine life [13]. Hence, this goal is dedicated to protect marine and coastal life. The first target is to be fulfilled by 2025, to reduce marine pollution from land-based activities. Ocean acidification is a new age challenge, so the next target under this goal is to take measures to increase scientific cooperation to reduce ocean acidification. Overfishing/unregulated fishing has to be controlled to maintain bio-diversity.

SDG 15: Life on Land

For sustainable living, it is necessary to protect the life on land. Goal 15 is dedicated for the protection of deserts, forests, mountain and wetlands. Conservation and restoration of terrestrial and inland freshwater, forests, mountains and dry lands is another target [14]. By 2030, it is important to restore degraded land and soil to reduce the frequency of drought and floods. In order to save the life of different species, it is equally important to restore and conservation mountain ecosystems. End poaching and trafficking is an important target under this goal. To protect the sustainable lifestyle, it is equally important to protect species of flora and fauna. Nations has to take strict actions for illegal trade of wildlife products.

SDG 16: Peace and Justice Strong Institutions

To make world a better place to live, it is necessary to increase the harmony and peace, for this, it is required to decrease violence in all forms. Abuse in any form is dangerous for the physical and mental growth of any human being. This goal is targeted to provide protection to all vulnerable groups from abuse/exploitation or torture. Information technology can make wonders in it. If transparency can be increase, corruption/bribery can be reduced drastically to protect the right of all vulnerable communities. The next target under this goal is to provide public information to all to increase national and international cooperation. It will also help to combat against terrorism and cross-border crime.

SDG 17: Partnership to Achieve the Goal
To achieve the ambitious SDGs, it is necessary to utilize the world resources in optimum way. For this international cooperation, foreign investment can become a catalyst. Strong bonding with different nations can bring the production cost at its lowest. Knowledge sharing and resource mobility can be possible when there is harmony in international relations. Developed counties can help the developing countries in developing with environment safe technologies. The United Nation has given 19 targets for SDG 17. These 17 SDGs have smaller targets within. Total 169 targets need to be focus to in these 17 goals. Each SDGs has high linkage with each other. No single goal can be achieved in isolation. It is observed that all actions in one area have high impact on the nearby areas.

Conclusion

The basic principle of SDG, which is achieving these 17 goals with equality needs high monitoring. Without the help of technology, it is not only possible to monitor the growth in individual goal. Almost every parameter of development is well taken care while framing the SDGs. The ill effects of environment degradation cannot be limit to single nation. If there is problem within the boundaries of one nation, it will affect the world environment too. The rich nations can afford to have better technology to cut down the emission and thus protect the negative effects on the life of their citizens but for a poor nation, it is difficult to create enough awareness among people to sensitize toward environment. Replacing the technology in production is also a difficult task. Here comes the role of artificial intelligence and deep technologies. It is proven that different deep technologies can make wonders in health care or save from natural calamities by issuing timely warnings and many others. Artificial intelligence (AI) can help to achieve different goals with transparency and cost effectiveness. This way the benefits can seep to lower stratum of society with more effectiveness. It is already proven that these computing technologies have high role in increasing production efficiencies and positively affect global productivity [15]. AI can be highly effective in optimal utilization of natural resources. For example, water is such a natural resource which gets recharge by the nature itself still it becomes a scarce resource. Water is so important that it can force people to migrate and thus change the entire demography of a region. With the help of AI, the forecasting of this resource can be done. The best part is timely measures can be taken to save the mankind from the acute water shortage. Indian agriculture is basically monsoon dependent, if water shortage can be predictive and communicated to the farmers in timely manner, then sowing pattern can be alter accordingly. Another common issue with low-middle income countries is water contamination. Water-related diseases are highly common in least developed countries. 2 billion people at global level forced to use a drinking water contaminated with feces (WHO 2019).[3] In all such scenario, AI can be used to track the quality of the drinking water. AI techniques can also be used in marinating the water level at reservoir in dam construction [16]. River management and maintenance of water quality of river can also be monitored through AI. Moreover, all these AI techniques are highly cost-effective. In this way, AI can be used to monitor SDG 3, SDG 6, SDG 11 and SDG 15.

Equality and inclusion are two pillars of SDG, both of these can be achieved with the help of computing technologies [17]. The cost of education can be drastically controlled through deep technologies. Education can be reached to everyone who wants to learn and that's too with minimum cost. Another hidden principle of SDG is to control negative externalities, with the assimilation and analyzing the data, it become easier to control the externalities [18]. For a low-middle income country, it is highly ambitious to achieve all SDG with regional equality. With regular monitoring of data on different SDG, it is possible to gradually increase the benchmark.

Future Scope

Deep technologies can help in identifying the local problems before the time of crisis. Even the possible solution can be suggested to communities in cost-effective manner. Deep technology can help communities and regions by issuing warning before any natural calamity. Deep technologies are such engineering innovations which can provide easy solution to complex issues like climate change, chronic diseases, production of clean energy and many others. It is different from traditional technology as it has a key focus to provide solution to local problems. Such technologies like robotics, vision and speech algorithms are already contributing in production in many developed countries but in developing countries, there is a need to apply these technologies to find the solution to different regional problems. Developing countries usually have huge untapped potential in term of natural resources. Different deep technologies can be effectively utilized to find such resources. With the help of these technologies, it is possible to divide bigger area into smaller regions to attack similar kind of problems. Big data can also help in maintain transparency, safety and ethical standards. This can be used for equitable resource allocation so that poverty can be reduced. Totally, different deep technologies can tackle each SDG in efficient manner. Without such scientific innovation, it is difficult to achieve such ambitious goals in time framed manner.

---

[3] https://www.who.int/news-room/fact-sheets/detail/drinking-water.

References

1. Ferri, N.: United nations general assembly. Int. J. Mar. Coast. Law **25**(2), 271–287 (2010). https://doi.org/10.1163/157180910X12665776638740
2. Todaro, M.P., Smith, S.C.: Economic Development: The Addison-Wesley Series in Economics (2015)
3. UNDP: Human Development Report 2019: Beyond Income, Beyond Averages, Beyond Today, United Nations Development Program (2019)
4. Nayyar, D.: In: R. Findlay (ed.) Catch Up: Developing Countries in the World Economy. Oxford University Press (2013)
5. Johnston, R.B.: Arsenic and the 2030 agenda for sustainable development. In: Arsenic Research and Global Sustainability—Proceedings of the 6th International Congress on Arsenic in the Environment, AS 2016, pp. 12–14 (2016). https://doi.org/10.1201/b20466-7
6. Food and Agriculture Organisation of the United Nations: Food Security and Nutrition in the World the State of Transforming Food Systems for Affordable Healthy Diets, the State of the World (2020). Available at:https://doi.org/10.4060/ca9692en
7. Ferguson, T., et al.: SDG4—Quality Education (2018). https://doi.org/10.1108/978178769 4231
8. UNESCO: Discussion on SDG 4—Quality education. High-Level Political Forum on Sustainable Development, p. 14 (2019). Available at: https://sustainabledevelopment.un.org/content/documents/23669BN_SDG4.pdf
9. Aid, W.: The Interlinkages Between Water, Sanitation and Hygiene (WASH) and Nutrition, vol. 2 (2019)
10. United Nation-HLPF: 2018 review of SDGs implementation: SDG 11—Make cities and human settlements inclusive, safe, resilient and sustainable. High-Level Political Forum on Sustainable Development, pp. 1–11 (2018)
11. Le Blanc, D., UN Environment Programme: Issue brief SDG 12: Ensuring Sustainable Consumption and Production Patterns. Dept. Econ. Soc. Aff. **1**(141), 1–17 (2017). Available at: https://wedocs.unep.org/bitstream/handle/20.500.11822/25764/SDG12_Brief.pdf?sequence=1&isAllowed=y%0A. http://www.un.org/sustainabledevelopment/sustainable-consumption-production/%0A. http://www.un.org/esa/desa/papers/2015/wp141_2015.pdf
12. Masson-Delmotte, V., Zhai, P., Pörtner, H.-O., Roberts, D., Skea, J., P.R.S., Pirani, A., Moufouma-Okia, W., Péan, C., Pidcock, R., Connors, S., Matthews, J.B.R., Chen, Y., Zhou, X., Gomis, M.I., Lonnoy, E., T. M., Tignor, M., T. W. (eds.): (1981) IPCC 2018. Nature**291**(5813), 285. https://doi.org/10.1038/291285a0
13. Cicin-Sain, B.: Conserve and sustainably use the oceans, seas and marine resources for sustainable development. UN Chron. **51**(4), 32–33 (2015). https://doi.org/10.18356/8fcfd5 a1-en
14. Broady, P.A.: Life on land. Exploring Last Cont. pp. 201–228 (2015). https://doi.org/10.1007/978-3-319-18947-5_11
15. Toro Suarez, L.Y.: The Economics of Artificial Intelligence (2015)
16. Mehmood, H., Liao, D., Mahadeo, K.: A review of artificial intelligence applications to achieve water-related sustainable development goals. In: 2020 IEEE/ITU International Conference on Artificial Intelligence for Good, AI4G 2020. Institute of Electrical and Electronics Engineers Inc., pp. 135–141 (2020). https://doi.org/10.1109/AI4G50087.2020.9311018
17. Beder, S.: Recommended Citation Recommended Citation Beder, Sharon, The Role of Technology in Sustainable Development (2000)
18. Fisher, J., et al.: Debiasing word embedding. In: 30th Conference on Neural Information Processing Systems, (NIPS 2016), pp. 1–9 (2020). Available at: https://code.google.com/archive/p/word2vec/

Chapter 10
Healthcare 4P: Systematic Review of Applications of Decentralized Trust Using Blockchain Technology

Deepika Sachdev, Shailendra Kumar Pokhriyal, Sylesh Nechully, and Sai Shrinvas Sundaram

Introduction

In the past few years, widespread technology adoption has introduced the concept of 4P medicine-predictive, preventive, personalized, and participatory. Accessibility of the Internet, IoT, and connected devices have empowered doctors, patients, and researchers globally to access reliable information to enable advancement in health care services. Healthcare 4P is enabling the fourth industrial revolution by ensuring that health care is automated, advanced, and reliable. Availability of verified and trusted data is imperative for the success of Healthcare 4P as mentioned by Azarm-Daigle et al. [1].

Blockchain is an upcoming technology which enables trust, automation and decentralization across all stakeholders. However, it also has its known challenges of scalability, regulatory compliance, and operational overheads as mentioned in Abadi et al. [2]. The objective of this book chapter is to analyse applicability as well as challenges of Blockchain in information sharing in the healthcare sector across key use cases of clinical data availability for research, sharing of health care information by patients and pay for performance.

D. Sachdev (✉)
Application and Analytics, South East Asia Nokia Software, Singapore, Singapore
e-mail: deepikaa.sachdev@gmail.com

S. K. Pokhriyal
School of Business, Himalayiya University, Doiwala, India

S. Nechully
Haris Al Afaq LLC, Abudhabi, United Arab Emirates

S. S. Sundaram
Aptitude Global, Hyderabad, India

As explained by the World Economic Forum society today is undergoing the fourth industrial revolution, in which humanity and technology combine to form "cyber-physical systems," and Blockchain will play an important factor in this [3]. The key challenges in healthcare industry today include information security, availability of trusted and quality data for harnessing Advanced Health Technology and Transparency. While the growth of data analytics and large-scale data availability can accelerate research in healthcare industry, it is seriously challenged due to the availability of quality healthcare data. Blockchain with its key tenants of immutability, decentralization and disintermediation will provide unique models for Health Care by enabling Information Sharing in a trusted, secure and efficient manner between healthcare providers and patients as identified by Pilkington [4]. This will help unlock information interoperability to bring financial as well as operational effectiveness in health care. It will help to overcome the challenges of trusted data, centralized platform for data access as well as owner-controlled data in the health care industry. This will bring rapid improvement in Research for Medical Sciences. With the advances in technology, there are multiple Blockchain frameworks available to cater to the needs of healthcare industry. This chapter reviews issues with the current Health Care Industry. This paper identifies industry challenges based on use cases. It performs an analysis of Blockchain frameworks in the industry based on known industry challenges. The outcome is the creation of a repeatable matrix for healthcare applications using Blockchain. It also provides recommendations on Blockchain roadmap features to enable trust and interoperability amongst healthcare institutions.

Analysis of Blockchain Applications in Removing Barriers for Information Sharing Barriers in Healthcare Sector

Research Methodology

This book chapter uses grounded theory and framework analysis technique for verifying the objectives since it has a pre-defined set of issues and responders.

Grounded theory helps to generate a theory from the qualitative data collected which has been analysed systematically. As the name signifies, the theory is grounded in the data collected by the researcher. Grounded theory can help to identify the patterns of relationships hidden in the data and transcripts.

Framework analysis is a qualitative technique which is used to get feedback from the users who implement the policies. Framework analysis provides a "Systematic Approach" for qualitative data analysis. It scientifically helps to reduce or summarize interview transcripts to rows and columns of framework matrix.

Analysis of Existing Barriers in Healthcare Vertical

The health information national trends survey data highlights the impact of digitization on health care with patients increasingly researching online information before approaching a health care professional as reported by Hesse et al. [5]. In addition, Blockchain research is getting increasingly popular in PubMed journals with approximately 40 papers focussed on Blockchain to introduce trust and decentralization in information sharing in health care as highlighted by Wilson et al. [6].

Berwick et al. [7] have aptly coined the triple aim of successful health care as an optimized balance between care, health and cost. This will be achieved by a combination of 5G, IoT, Blockchain and artificial intelligence. As per Deloitte report [8], IoMT (Health Care Monitoring and Diagnosis via IoT Devices) which was reported at $14.9 billion in 2017 is expected to increase to $52.2 billion by 2022. Connected devices will play a critical role to play in helping to improve standards of care and cost optimization to enable the transformation to value based care. To accelerate the same, analytics in healthcare industry is growing at an exponential rate of 27.3% as per Onik et al. and is expected to grow up to 29.84 Billion USD by 2022 [9].

By application of a combination of Open, Axial and Selective Coding, this Book Chapter analyses the current Blockchain Based applications in Health Vertical. The objective is to identify critical barriers in Health Care based on Industry use Cases (Fig. 10.1).

Aggregated Clinical/Administrative Data Availability for Research: The health care industry has transitioned from paper to online records, however, the exchange of data across healthcare providers is still limited [3]. Patient healthcare information is generated in various healthcare institutes and is owned by healthcare plan provider. This information is stored in siloes each in proprietary formats and security protocols. This becomes a barrier for information and payment sharing across patients, providers, insurance agencies and payers.

As per research by Brailer et al. [10] interoperability is a primary prerequisite for the economy to derive social and economic benefits from electronic medical records. Further as explained by Kruse [12] without interoperability, the silos which exist in the Industry will be further strengthened due to healthcare providers gaining exclusive rights over the patients information.

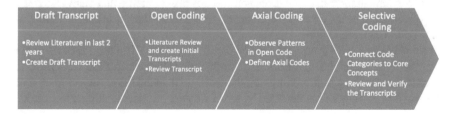

Fig. 10.1 Creation of conceptual lens

As explained by Taylor et al. [8] Medical Technology related (MedTech) companies fabricate more than five hundred thousand varied types of medical devices, ranging from wearable medical instruments, stationary medical devices and implants. To optimize the same Kruse et al. [12] analyses that the availability of standardized data structures, access security, as well as trusted storage will be key to ensure that the connected device data improves big data analytics in healthcare.

Security of Patient Data: Health Insurance Portability and Accountability Act (HIPPA) has pre-defined regulations governing the privacy of patient's data. Patient medical records have to be protected from modifications or security breaches as explained by Kumar et al. [13] It is important to highlight that patient information must be shareable and accessible to concerned parties including insurance providers.

However, malware attacks on healthcare records are quite common. Pilkington details that [4] in May 2017 thousands of British patients had delayed treatment since GPs and other critical services across NHS were impacted by malicious software by a virus created by US-based cyber warfare agents. Approximately 30 health service organizations had malware attacks, and many more were temporarily shut down as a precautionary measure.

Increasing data with connected devices should be a key driver for enabling patient healthcare information portability across stakeholders. The important considerations to make this successful are:

- Interoperable network across various healthcare providers that is used for validating healthcare information access control [9]
- As explained by Bram et al. in enabling of micropayments using automated protocols for access of healthcare literature and triggering of revenue sharing.

As mentioned by Gellman et al. [14] a key differentiator between data requirements of regulated and non-regulated organizations is the accuracy levels. Accuracy levels for marketeers can be lower as against those of healthcare providers. Hence, security requirements vary across business requirements and use cases.

Pay4Performance (P4P):

As explained by Kovacs et al. [15] Pay for performance (P4P) schemes are based on financial incentives to health care workers or agencies subject to achievement of pre-defined and agreed key performance indicators (KPI). Research on P4P schemes highlights that success outcome is based on performance plan definition, process-based implementation and execution. Another key observation is that many institute's fail to publish metrics for KPI's and measurement techniques resulting in failure of successful implementation.

As explained by Ayvaci [16] a key driver to enable P4P-based models is the availability of digital health information exchanges (HIE) which accurately provide information tracking points. However, the validity of such HIE is subject to trust of the provider and can be manipulated by the healthcare provider to tamper the results.

Insights of Blockchain Type for Healthcare Sector

As per a study conducted by IBM, approximately 16% of healthcare professionals are considering including Blockchain in their development roadmap. The important characteristics of Blockchain for the healthcare vertical are:

Transparency and Immutability: The key characteristic of Blockchain data is its immutability which implies that any and any updates to information, irrespective of off chain or on chain data storage is traceable and auditable to all [15]. Hence, Blockchain data storage ensures that information shared across organizations is visible, auditable and transparent to all as researched by Liang et al. [17].

Decentralization: Yang et al. analyse that [18] through decentralized information storage it ensures that no single institute controls the health care information. Information is available democratically since there is no single entity which owns the Blockchain [19].

Scalability: Kumar establishes that [13] one of the potential challenges in Blockchain healthcare is scalability since the cost trade off between available compute capability as against data storage of medical records could limit the potential of Blockchain for healthcare [19].

Interoperability: Current Blockchain platforms are not based on uniform protocols and hence have limited interoperability as verified by Siyal et al. [20]. Further as highlighted by Deepika [21] there are limitations due to proprietary formats across various providers.

Below is a detailed review of current Blockchain applications in healthcare vertical across industry applications identified based on literature study.

Clinical Data for Research

Guard Time: Is a UAE-based healthcare provider which is pioneering Blockchain application in health care. It works on online and offline mode on the KSI Blockchain which operates on proof of authority algorithm. It is a hybrid Blockchain which is energy efficient but limited decentralization since it is controlled by a limited number of actors. Each validating node publishes its blocks which are then validated by other nodes who have the power to reject. As highlighted by Matthias Mettler it is used by Govt. of Iceland, Estonia [22] and Hungary to maintain vaccination information for:

Proof of Vaccination: Similar to the yellow card this maintains vaccination information [23].

Priority Management Eligibility: To help the Govt. to have a fair and accountable process for priority management.

Monitor Uptake Amongst Population: Geographical monitoring of vaccine to ensure reduction in transmission.

Patient Information Security

Mediledger: In partnership with Genentech and Abbvie, Pfizer has piloted a Blockchain project coined as Mediledger. It uses a Closed/Private Blockchain to keep patient and drug traceability as recorded by Epiphaniou et al. [23]. Counterfeit drug production is stopped at the source by giving access rights only to manufacturers for serial number and product id association. Zero-knowledge proofs are used to ensure data privacy across organizations as explained by Beck et al. [24]. It ensures strong privacy and security arrangements by adopting key principles of data minimisation [25], similar to the Education Sector as explained by Deepika et al. [26]. In addition patients are given control of their information including viewing rights to see who has accessed their information [1].

MedRec [27]: A Blockchain pilot project led by the MIT Media Lab and Beth Israel Deaconess Medical Centre is addressing interoperability in Medical Data. MedRec operates in Offline storage mode wherein only the hash of the prescription is stored on Blockchain [27]. The patient has full access of the information and visibility into people accessing the information. MedRec is based on Ethereum. However, like typical Blockchain-based systems, the limitation exists that the user must trust the information published into the Blockchain since there is integrated verification system with external 3rd parties as explained by Esposito et al. [28].

Vaccination Status Tracking Through Blockchain

The COVID-19 situation has accelerated the need for vaccination information to be stored in a way that it can be provided to airline travel companies, border control and various access venues without a risk of fraud [4]. This requires that vaccination doses are tracked and the vaccination certificate that is issued is stored securely and is an immutable record that can be verified with ease [29].

The below picture shows the various government and international organizations using Blockchain for vaccination certification (Table 10.1).

Even though COVID-19 has accelerated the need for such solutions, these can be used to maintain vaccination records for individuals from birth enabling global movement and tracking of an individual vaccination status and ensuring these are kept current, for e.g. smart contracts that trigger reminders to parents to schedule appointments and book vaccination slots. Verified vaccination status can then be maintained on the Blockchain for the individual so that at any given point in time or in case of emergency their vaccine status is clearly understood and appropriate action

Table 10.1 Organization using Blockchain for vaccination tracking

Company name	Underlying technology	Used in solution
Algorand	Open source public Blockchain	VitalPass used in Colombia and plans to extend across Latin America
Evernym	R3/CuLedger/Hyperledger	IATA travel pass solution
Accredify	Hash is stored on the Blockchain	Digital health passport to store swab results and vaccination certificates
VXPass	RelayX and Bitcoin Computer	Verification of vaccination credentials through integration of various players in the healthcare provider ecosystem

can be taken [29]. This can also help in ensuring regional requirements for vaccination especially for children are automatically verified and validated and appropriate reminders triggered through a smart contract [3].

Interoperability will be a key requirement as individuals and families move across countries for leisure, business and permanent or temporary relocation with potentially both private and public Blockchain solutions being in place in various geographies [20].

Particularly for underprivileged children having a mechanism to track vaccination and creating triggers for subsequent shots can reduce child mortality and ensure vaccinations are not missed and there is a single source of truth and medical record.

The above can further support the indicators for United Nations Sustainable Development Goals (SDG) 3.b.1 with respect to "Proportion of the population with access to affordable medicines and vaccines on a sustainable basis" and enable nations to track vaccinations and report in a transparent and trustworthy manner.

Early Warning Sign for Epidemic

Once health records are placed on the Blockchain, through the use of smart contracts it can be possible to start identifying early warning signs of epidemics, for e.g. smart contract that monitors/reviews anonymized patient data for key indicators, for e.g. fever, cough, etc., and identifies commonalities in an area and triggers alerts to health authorities to investigate the data further [6]. Further work needs to be done to ensure the patient privacy vs the needs of the broader community are balanced and aligned [9].

Monetisation of Patient Data

There are strict rules for governing the use of patient data in most countries with anonymised data sets being used by hospitals or other health care providers. With the level of granular control possible with Blockchain, patients could potentially control the data that is shared and monetize the data through micro payments handled

Fig. 10.2 Trust
requirements for health care

through crypto currencies or tokens [30]. This allows the patient to be in total control
of their data and disclose that to the extent they are comfortable with [29].

Summarized View of Barriers and Applications of Standard Blockchain Types for Healthcare Vertical

See Fig. 10.2 and Table 10.2.

Analysis of Blockchain Models for Healthcare Sector

Summary of Conceptual Lens Based on Literature Review

The conceptual lens is based on six key themes analysed across three health care
use cases verified across nine key node items. Each theme consists of mutually
exclusive and collectively exhaustive characteristics. Access control is critical to
ensure patient data is available only to legitimate parties. Business rules is required
for verification as well as payment automation. Decentralization ensures that data is
owned by a single owner. Immutability brings the trust by ensuring that data has not

Table 10.2 Challenges and benefits of standard Blockchain types

Blockchain category	Blockchain provider	Use case	Applications	Access control	Volumetric	Barriers
Public permissioned	MediLedger [23]		ZK proofs used for drug ownership validation	Per missioned reads	Stores records of patients	Interoperability with other Blockchains
Private	Guard Time [31]	Issuing and validating health certificates that are tamper-resistant and permanent	It allows for more flexible API-based integration compare to other Blockchain solutions	Permissioned read and write	KSI ensures volume limitations are removed	Source data verification
Public	MedRec [19]	Patient health care information with access control	Democratize copyright due to a truly decentralized platform	Public access	Imposed by ethereum	Transaction volume supported by system

Table 10.3 Matrix coding for data analysis

Use case/theme	Pay4Performance	Clinical data for research	Patient data security
Access control	6	8	8
Business rules	6	2	4
Decentralization	4	3	2
Immutability	9	9	7
Interoperability	2	3	2
Technology use barrier	5	5	5

been tampered [32]. Interoperability is critical to ensure that data can be exchanged across various health care providers. And, most importantly the technology should be easy to operate by the masses (Table 10.3).

Initial Interview Protocol

Attached is the initial interview protocol used for conducting survey (Table 10.4).

Challenges in Information Sharing

The above conceptual lens has been verified using a random survey conducted from May to August 2020 for collecting challenges faced by more than hundred experts in the healthcare sector for P4P, clinical data for research and patient information security and usage of Blockchain using selected questions reviewed by Blockchain and healthcare domain experts. The responders resume was highlighted by doctors and patients to ensure that the survey provides appropriate profiles in both the domains (Table 10.5).

Based on the survey conducted above from a section of about hundred responders it can be ascertained with seventy five percent confidence that following is a summary of Blockchain requirements for healthcare vertical challenges across various use case (Table 10.6).

Final Interview Protocol Based on Survey Findings

See Table 10.7.

Table 10.4   Interview protocol for survey

Information categories
What are the categories of information you share across organizations?
What process do you follow for information sharing?
Technical usage barriers
Do you feel that more information exchange across organizations will bring operational efficiencies?
Have you ever faced a scenario where information from your organization has been plagiarized by others without getting consent from your organization?
What are the barriers that may come in digitalization of information sharing system in your organization? Example lack of infrastructure/unskilled employee/high maintenance cost/cost of digitalization is high?
Information data storage and volumetrics
What is the volume of information shared in your hospital?
How is different set of data is maintained in your hospital?
Does the size of organization (large/small) contribute to how efficiently information is being shared? Example large organisation require more time in information sharing from one department to other/in small organisation lack of sufficient people and infrastructure causes delay in information sharing ?
Immutability and decentralization
What is the impact if information shared by you is modified by third party organizations? E.g., if student tampers with degree certificate or if course content is modified
Do you get feedback if you information is modified by third party?
Are there some organizations whom you would allow to modify the information while others cannot i.e., some trusted parties?
How is data updated when any correction is required? Time taken in the whole process?
What is the impact of interoperability and regulatory in data sharing?
Was there a scenario where you could not exchange information with other organization since the data formats did not match?
Has there been a case where you could not receive information from other organization due to regulatory challenge?
Cost of information sharing
How many people are involved in information sharing across your department?
Total average monthly salary for people involved in information sharing?
What is the total volume of data shared in the organization?
In your opinion how can this cost be reduced? And how would it effect total revenue generated by organization?
Data security/access control/trust/RBAC/identity management/audit trail/ownership
How is data secured in your organization?
What are the challenges faced in maintaining a secure system?
How many times leak of information being reported? What action were taken to prevent it?

(continued)

Table 10.4 (continued)

What is the importance of trust in information sharing with partners?
What is the process of giving unique identity to entities in the organization?
Do you see a need for role based access control in information sharing?
Business rules/interoperability/IP protection—health care
Does the patient have to repeat the examination when he moves from one hospital to the other?
Is there a way to review the data entered by drug companies into the database?
Is there sufficient transparency in health care information available across organizations?
Perceived benefits of information sharing on Blockchain
Has the organization considered information sharing using Blockchain and are there are known challenges (for e.g.: lack of business vision for information sharing lack of trust in technological systems of business sharing)?
Is there a perceived use case where it is believed that Blockchain will bring significant benefits to the organization?
Do you see operational savings by introducing Blockchain for information sharing? (for e.g., reduced human effort, reduced paperwork, faster time to market)?
Do you see increased transparency and trust with partners by introducing Blockchain for information sharing?
Do you see cost savings by introducing Blockchain for information sharing? (for e.g.: automation of processes and reduced paperwork and human effort)
Do you see new business opportunities by introducing Blockchain for information sharing? (for e.g.: information monetization, new university partnerships etc.)
Do you see increase in stakeholder satisfaction by introducing Blockchain for information sharing?
Does the data need to be stored on chain/off chain for security?

Interview Responses

The summary of analysis from the above survey has been further shared with Industry experts including Senior Professor, GB Pant Hospital (New Delhi), Head of Innovation, Celcom, Malaysia), Senior VP, Siemens Health (Bangalore) and Director Philips Innovation Campus (Bangalore). Key observations are:

- There is insufficient trust across healthcare organizations and information modification without owner permission can be done. It is also observed that lack of IT awareness amongst health care department workers leads to increased challenges
- Blockchain has the benefit of an immutable database. However, it has a limitation on the size of data stored due to block limitations and performance. Typical healthcare data comprising of electronic imaging records require large storage requirements and cannot be stored in its entirety on the Blockchain. This necessitates the use of hybrid Blockchains which enable storage of large data volumes. However, this introduces complexity which is hard to justify. Beyond being an immutable database, blockchain's USP is the trust factor. Key use cases for Blockchain are

Table 10.5 Information access in healthcare

Survey theme/question	Significance	Observation
IS-RQ1: what is the profile of the respondents	Characterizing based on profile Information we try to understand the distribution of expertise, the strength (and durations) of their attachments to current jobs, and seniority in the organization. There is a common concept that information sharing is dominated by younger respondents who are open to newer ideas. Hence, research questions were focussed on age, demographics, profession and seniority	• Most of the respondents were between the age group 20–30 and 70 and above • Only about 7.4% of respondents were getting their prescriptions online • This 7.4% of respondents belonged to each age group
IS-RQ2 what are the primary objectives of the respondents in information sharing	To understand important use cases and challenges in healthcare	• Verification of records due to lack of digital information made available by doctors is one of the important issues faced by more than 67% responders resulting operational issues and increased time
IS-RQ3: **Immutable**: need for immutable information share in inter organization and external organization information sharing	Research questions were focussed to understand if data immutability is critical and important for the organization since this is an important trait of Blockchain	• About 80% of respondents did not prefer sharing data with any external organization or any public source from where the data could be stolen • According to the responses gathered 7.4% of respondents felt that their health information is being illegally used
IS-RQ4: role-based access control (RBAC) for information exchange across departments	Research queries were directed to understand need for trust and security based on the organization needs	• About 7.4% of the respondents who were getting their prescriptions digitally, their data was accessed by the doctors of the same organization, and no external organization doctors were allowed to access their data • EHR data other than the regular health check-up was kept confidential and was only accessed by limited people

(continued)

Table 10.5 (continued)

Survey theme/question	Significance	Observation
IS-RQ5: **interoperable**: average number of times Information is shared in inter organization and external organization information sharing	Research questions were targeted to get details of data exchange frequency across departments	• About 36.1% of respondents need to transfer the data from one health centre to another • About 91.3% of the 36.1% of respondents did this transfer of data in written format, i.e. the written prescription • About 5.8% of the 36.1% of the respondents transferred the data digitally • About 46.3% of respondents had to repeat their medical examinations when they had to switch from one hospital to another
BC-RQ1: **decentralized**: application in health care vertical to verify need of decentralization	Review with industry experts the requirement of Blockchain in the health care vertical	• Less awareness of Blockchain technology • More than 80% of the respondents found Blockchain safer and secured for storing and sharing their data

Table 10.6 Information sharing challenges for healthcare vertical

Use case	Immutability	Permission read/write	Decentralization	Inter-operability	Business rules	Technology barrier
PayForPerformance	High	Medium	High	Medium	High	Low
Clinical data for research	Medium	High	High	High	Medium	Medium
Patient information security	High	Low	Medium	High	Low	Low

Table 10.7 Final interview protocol

Information categories

What are the categories of Information you share across Organizations

What process do you follow for Information Sharing

Technical usage barriers

Have you ever faced a scenario where Information from your Organization has been plagiarized by others without getting consent from your organization?

What are the barriers that may come in digitalization of information sharing system in your organization? Example lack of infrastructure/unskilled employee/high maintenance cost/cost of digitalization is high

Information data storage and volumetrics

What is the volume of information shared

Does the size of organization (large/small) contribute to how efficiently information is being shared? Example large organisation require more time in information sharing from one department to other/in small organisation lack of sufficient people and infrastructure causes delay in information sharing

Immutability and decentralization

What is the impact if Information shared by you is modified by third party organizations? E.g., if student tampers with degree certificate or if course content is modified

What is the impact of regulatory in data sharing

Cost of information sharing

How many people are involved in information sharing across your department

Total average monthly salary for people involved in information sharing?

In your opinion how can this cost be reduced? And how would it effect total revenue generated by organization?

Access control

How is data secured in your organization?

What are the challenges faced in maintaining a secure system?

How many times leak of information being reported? What action were taken to prevent it?

What is the process of giving unique identity to entities in the organization?

Do you see a need for role based access control in information sharing

Business rules—health care specific queries

Does the patient have to repeat the examination when he moves from one hospital to the other

Is there sufficient transparency in health care information available across organizations

Information sharing on Blockchain

Has the organization considered information sharing using Blockchain and are there are known challenges (For e.g.: lack of business vision for information sharing lack of trust in technological systems of business sharing)

Is there a perceived use case where it is believed that Blockchain will bring significant benefits to the organization

(continued)

Table 10.7 (continued)

Do you see operational savings by introducing Blockchain for information sharing? (for e.g., reduced human effort, reduced paperwork, faster time to market)
Do you see new business opportunities by introducing Blockchain for information sharing? (for e.g.: information monetization, new university partnerships etc.)
What is the volume of data that needs to be stored in the Blockchain?
Does the data need to be stored on chain/off chain
Do you visualize the need to be shared with a trusted consortium or with any untrusted party globally?

focussed on areas where the involved parties do not trust human agency and need to remove the intermediary.

- P4P for health care is identified as a strong use case for Blockchain since it enables trust. P4P necessitates measuring of performance based on disease registers. Disease registers are the healthcare equivalent of ledgers. They record limited data sets such as blood pressure, heart rate on monthly visits and hence do not need high data storage or off-chain storage requirements. In addition, there are extensively documented rules sets for incentivization which can be coded using smart contracts. In addition, payments can be automated based on smart contract trigger rules.
- All that said, the trustworthiness of Blockchain technology does not preclude the necessity of audits in the real world. Registers maintained on Blockchain have to be verified by physical visits to the GP and meeting with some of the patients to verify the numbers recorded in the registers.
- Interoperable information across multiple Blockchains leads to serious concerns and leads to lack of information exchange and verification across departments.

Review of Challenges Across Blockchain Categories

This paper has analysed Six Key Barriers for existing healthcare sector information sharing across three industry models for Blockchain-based content. Below is a summary of identified gaps for roadmap development.

Blockchain Interoperability

It is important for healthcare vertical to exchange information across various organizations. Following interoperability issues exist:

- **Regulatory Protocols**: To be resolved at International and Country level to ensure improvement in Interoperability across Healthcare organizations.

- **Proprietary Protocols**: Data storage standards are unique to every application. They often cause limitations and time delays in data exchange across organizations. It is frequently seen that data exchange across health care organizations is limited due to lack of inter operable standards across organizations
- **Open API**: Like open banking initiatives, creating API for sharing information in standard formats across various applications with appropriate security controls will accelerate adoption.

Data Storage Limitations

One of the identified challenges in Blockchain healthcare is scalability due to cost of compute resources. Considering the volume of electronic health records, it would not be economical to store it on chain. There are options available for offline storage but that will add to security risks. In order to ensure that validated records are stored once and accessible through the Blockchain network from the underlying application storing the information, standards for medical record sharing needs to be defined and agreed by the various health care providers.

Verify Data Upload

Fraudulent institutes can inject invalid source data into the Blockchain. Due to the architecture of Blockchain there do not exist automatic verification processes to authenticate the source data which is published to the Blockchain for data storage. The data stored in the Blockchain is immutable, but existing Blockchain patterns do not have a separate authentication service for verifying the authenticity of the source data. This can lead to invalid master information in the system resulting in data corruption.

Validated Framework for Health Care

The below framework depicts the final conceptual lens for health care created based on interview response (Fig. 10.3).

Contribution to Literature

Multiple challenges inhibit information sharing in the industry. The information sharing needs vary based on the trust requirements of the industry. Blockchain provides a secure trusted solution for information sharing. This research paper contributes to literature as follows:

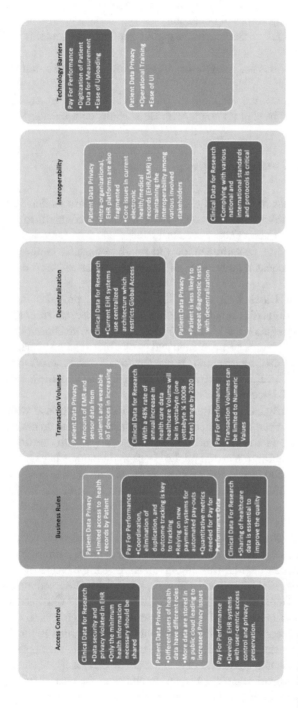

Fig. 10.3 Validated framework for health care

- A new "Trust Requirements Framework" created in health care vertical to enhance information sharing across organizations
- A Blockchain industry framework mapping application capability with "Trust Requirements"
- The framework will act as a source of reference for industry to baseline their trust requirements
- It will be also beneficial for industry Blockchain adoption by creating a cross metric analysis of trust requirement with Blockchain solution capability.

Conclusion and Future Scope

This book chapter has reviewed Blockchain industry standard models in relation to the challenges in healthcare sector to analyse the issues as well merits to ascertain the overall applicability. Each industry standard Blockchain patterns provides different degrees of immutability, decentralization and role-based access control. Below is a summary of the business requirements of the three use cases based on the survey (Table 10.8).

The analysis of Blockchain models in the industry provides the following summary across various Blockchain categories and their characteristics (Table 10.9).

Clinical data aggregation works across untrusted parties with large volumes of data share organizations. Hence, permissioned and public blockchain standards are more suitable for clinical data aggregation. However, for scenarios where the institutes decides to limit the level of decentralization, it could also become a candidate for private permissioned Blockchain patterns.

Patient data security has high requirements for decentralization as well as access control since the data can be provided by multiple institutes who do not have known trust. In addition, the data volumes are high since typically health care documents are volume intensive. Permissioned and public Blockchain models have higher degrees of decentralization and hence are applicable for patient data security protection. However, for scenarios where the Institutes decides to limit the level of decentralization, it could also become a candidate for private permissioned Blockchain patterns.

P4P has low data storage requirements since the data needed for verification data is limited. The need for decentralization and access control permissions is limited since it can be operated across trusted organizations for exchange of information. Moreover, the business rules are well defined and hence justifies medium complexity of business rules. There can be additional Health care institutes who need to be added to the consortium, hence the need for inter-operability is high. Based on the needs of this use case it becomes a candidate for a public permissioned Blockchain.

Based on this research study, cost model-based transaction analysis shall form scope for further research studies. Blockchain is in an early stage of adoption where costs for operationalization are not known. In addition, the cost of transactions for verticals includes intangible variables such as reputation and opportunity. Hence, at

Table 10.8 Use case requirement summary

Use case	Decentralization	Business rules	Access control	Transaction volumes	Inter-operability	Technology barrier
Aggregated clinical/administrative data availability for research	Very significant since need for access by untrusted parties	Medium complexity reduced entities in value chain	Medium significant since access control defined in rules for trusted parties	High volume accessible by all and large data volumes	High requirement need for accessible by multiple platforms	High barrier since used by hospital staff
Patient data privacy	Very significant since need for access by untrusted parties	Medium multi party value chain	Very significant since need for RBAC to control permission	High volume Accessible by all and large data volumes	High requirement accessible by multiple platforms	High barrier since used by hospital staff
Pay4Performance	Less significant since need for access by untrusted parties	Medium: rules restricted to pre-defined set	Medium significant since access control defined in rules for trusted parties	Low volume data volumes are low since data sets are small	High requirements accessible by multiple platforms	Medium barrier since used by skilled staff

Table 10.9 Blockchain model summary

Blockchain framework	Decentralization	Business rules	Access control	Transaction volumes	Interoperability	Technology barriers
Public permissionless	Very significant since no central authority	Medium complexity since change has to be transmitted to public nodes	Limited since public access	Supports low volumes due to on chain data storage	Medium since controlled by central regulatory	High since public keys and OS upgrade
Private permissionless	Medium significant since consortium owned but permissionless	High complexity since owned by consortium	Can control write access have based on user roles	Supports high volumes	Low since governed by private parties	Medium since private controlled
Public permissioned	Medium significant since public but permissioned	Medium complexity since change has to be transmitted to public nodes	Control write access based on user roles	Supports medium volumes	Medium since controlled by central regulatory	High since public keys and OS upgrade
Private permissioned	Low since consortium owned and permissioned	High complexity since owned by consortium	Control write access based on user roles	Supports high volumes	Low since governed by private parties	Medium since private controlled

this point it is not possible to create a direct cost correlation matrix of existing cost of implementation versus the actual cost of operationalization in Blockchain. This shall form part of future study.

References

1. Azarm-Daigle, M., Kuziemsky, C., Peyton, L.: A review of cross organizational healthcare data sharing. Procedia Comput. Sci. **63**, 425–432 (2015)
2. Abadi, J., Brunnermeier, M.: Blockchain Economics, Princet. U. Work. Pap. 1–53 (2018)
3. Ingraham, A., St. Clair, J.: The fourth industrial revolution of healthcare information technology: key business components to unlock the value of a blockchain-enabled solution. Blockchain Healthc. Today 1–4 (2020)
4. M. Pilkington, Can Blockchain Improve Healthcare Management? Consumer Medical Electronics and the IoMT, SSRN Electron. J. (2017).

5. Hesse, B.W., Nelson, D.E., Kreps, G.L., Croyle, R.T., Arora, N.K., Rimer, B.K., Viswanath, K.: Trust and sources of health information. Arch. Intern. Med. **165**, 2618 (2005)
6. Kamel Boulos, M.N., Wilson, J.T., Clauson, K.A.: Geospatial blockchain: promises, challenges, and scenarios in health and healthcare. Int. J. Health Geogr. **17** 1–10 (2018)
7. Berwick, D.M., Nolan, T.W., Whittington, J.: The triple aim: care, health, and cost. Health Aff. **27**, 759–769 (2008)
8. Taylor, K., Sanghera, A., Steedman, M., Thaxter, M.: Medtech and the internet of medical things: how connected medical devices are transforming health care, Deloitte (2018)
9. Onik, M.M.H., Aich, S., Yang, J., Kim, C.-S., Kim, H.-C.: Blockchain in Healthcare: Challenges and Solutions. Elsevier Inc. (2019)
10. Brailer, D.J.: Interoperability: the key to the future health care system. Health Aff. (Millwood). Suppl Web 19–21 (2005)
11. Khezr, S., Moniruzzaman, M., Yassine, A., Benlamri, R.: Blockchain technology in healthcare: a comprehensive review and directions for future research. Appl. Sci. **9**, 1–28 (2019)
12. Kruse, C.S., Goswamy, R., Raval, Y., Marawi, S.: Challenges and opportunities of big data in health care: a systematic review. JMIR Med. Inf. **4**, e38 (2016)
13. Kumar, T., Ramani, V., Ahmad, I., Braeken, A., Harjula, E., Ylianttila, M.: Blockchain utilization in healthcare: Key requirements and challenges. In: 2018 IEEE 20th International Conference on e-Health Networking, Applications and Services Heal. 2018, pp. 1–7 (2018)
14. Gellman, R.: Health Information Privacy Beyond HIPAA: A 2018 Environmental Scan of Major Trends and Challenges (2017)
15. Kovacs, R.J., Powell-Jackson, T., Kristensen, S.R., Singh, N., Borghi, J.: How are pay-for-performance schemes in healthcare designed in low- And middle-income countries? Typology and systematic literature review. BMC Health Serv. Res. **20**, 1–14 (2020)
16. Ayvaci, M., Cavusoglu, H., Kim, Y., Raghunathan, S.: Payment mechanisms, incentives for adoption and value of health-information exchanges (HIEs). SSRN Electron. J. 1–48 (2017)
17. Liang, X., Zhao, J., Shetty, S., Liu, J., Li, D.: Integrating blockchain for data sharing and collaboration in mobile healthcare applications. In: International Symposium on Personal, Indoor and Mobile Radio Communications PIMRC, pp. 1–5 (2018)
18. Yang, Z., Zhang, L.: Information sharing in supply chain: a review. J. Digit. Inf. Manag. **11**, 125–130 (2013)
19. Labazova, O., Dehling, T., Sunyaev, A.: From hype to reality: a taxonomy of blockchain applications. In: Proceedings of 52nd Hawaii International Conference on System Sciences (2019)
20. Siyal, A.A., Junejo, A.Z., Zawish, M., Ahmed, K., Khalil, A., Soursou, G.: Applications of blockchain technology in medicine and healthcare: challenges and future perspectives. Cryptography. **3**, 3 (2019)
21. Sachdev, D., Studies, E., Management, H.E., Studies, E., Studies, E.: Analyzing Blockchain Based Models For Digital Content Metadata **11**, 277–292 (2020)
22. Mettler, M.: Blockchain technology in healthcare: The revolution starts here. In: 2016 IEEE 18th International Conference on e-Health Networking, Applications and Services Heal. 2016, pp. 16–18 (2016)
23. Epiphaniou, G., Daly, H., Al-Khateeb, H.: Blockchain and healthcare. Adv. Sci. Technol. Secur. Appl. 1–29 (2019)
24. Beck, R., Avital, M., Rossi, M., Thatcher, J.B.: Blockchain technology in business and information systems research. Bus. Inf. Syst. Eng. **59**, 381–384 (2017)
25. Lemieux, V.L., Ph, D., Hofman, D., Batista, D.: Blockchain Technology & Recordkeeping Table of Contents (2019)
26. Choudhury, T.: Education 4.0: A Systematic Industrial Case Based Review of Barriers and Applications of Decentralized Trust Using Blockchain, vol. 57, pp. 882–893 (2020)
27. Azaria, A., Ekblaw, A., Vieira, T., Lippman, A.: MedRec: Using blockchain for medical data access and permission management. In: Proceedings of the 2nd International Conference on Open Big Data, OBD 2016, pp. 25–30 (2016)

28. Esposito, C., De Santis, A., Tortora, G., Chang, H., Choo, K.K.R.: Blockchain: a panacea for healthcare cloud-based data security and privacy? IEEE Cloud Comput. **5**, 31–37 (2018)
29. Agbo, C.C., Mahmoud, Q.H.: Blockchain in healthcare. Int. J. Healthc. Inf. Syst. Informatics. **15**, 82–97 (2020)
30. Radjenovic, Z.: The cost-saving role of blockchain technology as a data integrity tool: e-health scenario. KnE Soc. Sci. **2020**, 339–352 (2019)
31. Sullivan, C., Burger, E.: E-residency and blockchain, Comput. Law Secur. Rev. Int. J. Technol. Law Pract. 1–12 (2017)
32. Aggarwal, A.: Review of ownership based blockchain frameworks in government applications, Annu. J. Inst. Innov. Technol. Manag. **3**, 22–32 (2016)

Chapter 11
Implementation of an IoT-Based Water and Disaster Management System Using Hybrid Classification Approach

Abhishek Badholia, Anurag Sharma, Gurpreet Singh Chhabra, and Vijayant Verma

Introduction

Disaster management may be described as "the spectrum of actions aimed to maintain control over catastrophe and emergency situations and offer a framework for helping the individuals at risk to prevent or recover from the impact of the disaster." When natural or man-made disasters occur, disaster management is the discipline of dealing with and preventing hazards. To minimize or reduce the effects of catastrophes originating from hazards, disaster management is a continual process through which people, groups, and communities manage risks in an efficient manner.

Flood Disaster Detection System Based on IoT

Natural catastrophes are a concern for governments across the world. It is not uncommon for natural catastrophes to hit places where humans live. Nature- or man-made disasters are the most prevalent. Only 2% of the country's area can support a sixth of the world's population. A very rain storm in India in 2019 led to the death of at least 200 people and the displacement of millions. Kerala and Maharashtra suffered the most. It is common for floods to occur in coastal locations, along rivers and dams. In order to decrease exposure to flooding on a real-time basis, a real-time monitoring and warning system is needed. Disasters can be avoided if disasters are detected early enough. The three rain sensors and three water sensors are located in three separate locations, and a microcontroller with water sensors and rain sensors is also used in this system. A flood is predicted and an alert is sent to the appropriate authorities,

A. Badholia (✉) · A. Sharma · V. Verma
Department of Computer Science and Engineering, MATS University, Raipur, Chhattisgarh, India
e-mail: abhibad@gmail.com

G. S. Chhabra
Department of Computer Science and Engineering, Gandhi Institute of Technology and
Management, Visakhapatnam, Andhra Pradesh, India

© The Author(s), under exclusive license to Springer Nature Singapore Pte Ltd. 2023   157
V. Kadyan et al. (eds.), *Deep Learning Technologies for the Sustainable
Development Goals*, Advanced Technologies and Societal Change,
https://doi.org/10.1007/978-981-19-5723-9_11

and a warning is sent to surrounding communities via the Internet of Things (IoT). As well as calculating flood times, the system also estimates how long it will take for the water to reach the specified region. As opposed to rain sensors, which measure rainfall intensity as millimeters, water flow sensors monitor the amount of water in a water body. From a distant location, the full information may be accessed, giving the necessary information on the IoT platform Predicting floods is largely influenced by rainfall. There have been a number of machine learning methods tested for flood prediction models in the past. The chapter's major contribution is on the role of deep learning models. As a result of efforts to enhance flood prediction models, fewer people have been killed, disasters have occurred, and assets have been damaged. To anticipate the advent of floods, this chapter proposes a hybrid deep learning approach that uses rainfall data from past months to estimate the upcoming month's rainfall. Modeling requires no meteorological or geographic information of its hydrological and topographic characteristics. The study has found numerous flood prediction models, including radar systems, hydrographic analysis, a network of rain gauges and stream networks, as well as statistical models and time-series analysis predictions models. To train CDNN + ANN, DL models were employed. When it comes to monsoon seasons, rainfall prediction is regarded highly essential.

In this article, Big Data and the recommended CDNN + ANN are used to identify floods. Two phases of training and testing are involved. It starts out with a flood-related large data set that contains data on water flow and water level, together with information from the Rain Sensor and Humidity sensor (HM). For the remainder of the process, the Hadoop Distributed File System and Map Reduce technique were utilized. Following the normalization procedure, the characteristics are used to generate a function. The hybrid classifier uses the combined attribute function as an input and separates the flood detection into chances and no chances. This is when IoT values are used as input for the classification process. The rest of the procedure is identical to the training phase, including HDFS, preprocessing, and categorization of data. The remaining section of the chapter can be organized as follows, Section "Introduction" includes the description of flood disaster management and its importance. Section "Related Works" includes the related work concerning the IoT-based disaster management. Section "Problem Statement" depicts the problem statement. Section "Proposed Methodology" includes the overview of the procedure suggested. Section "Result and Discussion" contains the interpretation of the suggested procedure. Section "conclusion" concludes the work.

Related Works

How a wireless sensor network powered by the Internet of Things is being created and tested in the Mexican city of Colima-Villa de lvarez [1]. 3G and Wi-Fi networks are used to capture data on fluvial water level and soil moisture, which is then delivered in real-time to a server and a web application through IoT message queuing Telemetry Transport protocol. As part of the 2019 tropical cyclone season, three distinct tropical storms passed through the Colima region. Tropical storms, for example, have shown that the smart water network can collect real-time hydro meteorological data. Rani et al. [2] contains a flood detection technique that is both effective and adaptable, as well as an alarm system. Technological advancements such as machine learning (ML) are a benefit to the area of engineering since they are extremely effective in monitoring the normal and abnormal behaviors of any system. In this project, we will conduct a survey on flood concerns. Rainfall forecasting relies heavily on neural networks, which are extensively used and perform well [3]. Technology that can not only detect the water level, but can also measure the pace at which it rises and alert inhabitants to the danger. The waterfall paradigm is utilized in this project as a method. Data from the water sensor may be collected using a Raspberry Pi and sent to a GSM module to send an SMS alert to your smart phone. For example, it will be shown how the Raspberry Pi may be linked to a smart phone in order to send an alert. To guarantee that the system can give accurate and trustworthy data, it is evaluated in an experiment consisting of two separate environments. As an IoT-based project that supports the infrastructure of Cyber-Physical System, the project is in accordance with Industrial Revolution 4.0 [4]. An ultrasonic height detector, an Arduino Uno microcontroller, and a U-Blox Neo 6m GPS module and GSM module have been used to create a prototype flood monitoring system based on Google Maps. Based on Google Maps interface, the prototype's design delivers flood altitude data together with its location [5]. The chapter describes the design of a flood monitoring system using LoRa technology, as well as its installation and testing results. So that it may link multiple types of sensors without requiring substantial hardware changes to the proposed node architecture, the entire system is built from a modular perspective. Sensors and a microcontroller are used to save the data. The data is then analyzed and stored in an online structure where an alarm function is created if there's flooding [6]. An IoT-based smart flood monitoring and forecasting architecture is proposed based on the convergence of big data and high-performance computing (HPC). Hexagonal networks are used to deploy energy-efficient IoT devices effectively. They allow all flood-producing or flood-preventive characteristics to be recognized using big data and high-performance computing (HPC) approaches. The SVD algorithm is used to decrease the number of features. A technique called "K mean clustering" is employed to assess the current flooding and flood rating in any area, while the Holt-forecasting Winter's approach is utilized to predict the flood in any location [7]. In the event of a flood, send a text message warning. With the use of SMS, the authors intend to allow the relevant authorities to take the required action. A sensor is a device that detects and reacts to physical data. As early as 1981, Intel unveiled the 8051 microcontrollers. The microcontroller in this example has an 8-bit architecture. Two timers and 16 bits are included in the 40 DIP pins. There are two timers included in the package. It has four parallel 8-bit ports that may be programmed according to the needs. When a GSM modem, also known as a GSM module, establishes contact with another network, it does so by using GSM mobile technology. These switches are used in tanks both as level sensors and fluid control devices. In a condenser, electric energy is stored in a magnetic field. Because it has two terminals, it is an electronic gadget that's passive. When a current is running through a light-emitting diode (LED), it emits light. An acronym for LCD is Liquid Crystal Display (liquid crystal display). Many electrical gadgets utilize a tiny flat display device to show data. Signals are transformed from one logic or voltage domain to another by using a digital level shifter [8]. Artificial neural networks (ANN), long short-term memory (LSTM), and gated recurrent unit are some of the models we employ for flood forecasting (GRU). Kumar et al. [9] to depict a system that connects our arrangement with a real-time cloud that interacts with our mobile phones to check water levels Here, we use IoT to anticipate floods based on Arduino Uno's output and inform the appropriate authorities. In addition, the system predicts the time it would take for floodwaters to reach them and offers IVR, alerts to individuals so that they may be evacuated or manage their time accordingly. Khalid and Shafiai [10] outlines the role of the government's delivery system for flood victims in Malaysia before, during, and after the catastrophe. A technology-centered approach to flood control, forecasting, warning, and evacuation systems dominates the delivery system for flood management at the present time, with a heavy emphasis on the use of modern technologies. When a flood catastrophe strikes in Malaysia, the National Security Council (MKN) is in charge of managing the national disaster management system. It will also seek to discuss the sort of flood delivery system that has been employed in Malaysia and make similarities with delivery systems in other nations in this document [11]. Rainfall and river water levels to determine their temporal correlations for flood prediction study. When it comes to flood prediction, the IoT method is utilized to gather data from the sensors and communicate through Wi-Fi, while ANN is used to analyze the data. Wijekoon et al. [12] identifies a suitable, integrated software solution to the problem. These operations are performed in an integrated manner with the aid of the IoT technologies such as android, cloud, and windows applications [13]. GIS-based analysis and selection of evacuation routes that was both effective and safe from floods (GIS). Ghapar et al. [14] reviews the research on IoT for flood data management that has been done. As a result of this, the article suggested an IoT architecture for flood data management that may serve as the basis for the installation of IoT infrastructure that collects, transmits, and manages flood related data [15]. The Mamdani fuzzy inference system with rule-based reasoning to study flood events prediction. There are three parameters that determine the input and output of the model: population density, altitude, and rainfall. The membership functions for each variable are three to four.

Problem Statement

Globally, natural catastrophes lead to extensive devastation of property, as well as bodily injuries and even deaths. However, despite the fact that natural disasters are unpredictable, mitigation techniques can lessen the degree of the damage caused by these events. It is possible to decrease the effects of natural disasters by implementing effective flood protection techniques. It is therefore important to gather information from reputable sources. Risk assessment and risk management in the community require effective and timely data transmission. One of the most frequent natural dangers in Asia is flooding. Increasing human activity increases flood danger and impacts. It is important to have a flood prevention and reduction system that is integrated. As a result, geographical data integration, quick and accurate catastrophe information, and an effective detection tool for instant warning in the inundated regions from IoT DAAS, if they exist, may be used as an early warning system to limit possible damages. In disaster management, the problem is not a lack of technology or information. It is usually because the information is not easily accessible. Finding disaster recovery solutions requires a spirit of inquiry and a desire to utilize information technology efficiently to uncover and organize information, analyze it critically and correctly apply it to solve a problem. As a result, identifying, obtaining, and accessing data, as well as deciding, are important for implementing an effective response to the issue. Online information distribution is widely acknowledged as a benefit of Internet technology. Between service providers and users, it serves as a means of communication. Because of these developments there has been a significant increase in the number of Internet apps and services, including web-mapping services. Sensors, for example, are IoT service providers that give environmental information to the public. There are millions of people that utilize this service since it is quick and easy to use. Sensor information is provided by these IoT services apps. In typical settings, they are good at spotting regions or locating locations. Majority of them do not work as a real-time catastrophe warning system, but in the case of natural disasters such as flood alerts, they do not serve their function. The main directive for disaster management is to respond immediately to a flood occurrence. Prior to an emergency period, during an emergency period, and after an emergency period, this comprises ways for acquiring urgent data as well as timely communication in three stages. In the event of a flood disaster, sensors are a potential option. Detecting the flooded regions from sensor data is also essential.

Proposed Methodology

An urgent need exists for a system that can gather and monitor flood data in real-time, process it, and disseminate it in a timely manner. Developing an algorithm to detect inundated regions through an interactive system is one of the major roles that have yet to be determined. It is a vital component of the spatial information system, which may be used to build helpful applications in the case of a flood catastrophe. In addition, it increases the efficacy and efficiency of reacting to a flood application by improving access to sensor data.

The mechanism is depicted in Fig. 11.1. As a whole, the plan calls for setting up a technical infrastructure that will allow for rational decision-making, as well as implementing and developing algorithmic methods for the processing of sensors' data. IoT hardware and IoT platform are combined to create a flood monitoring system that can be watched from a remote place, as shown in Fig. 11.2. In the next stage of the system, DL models are used to predict rainfall. A. IoT interface block diagram: For example, a Raspberry Pi 3 with a Wi-Fi module, LED display with a

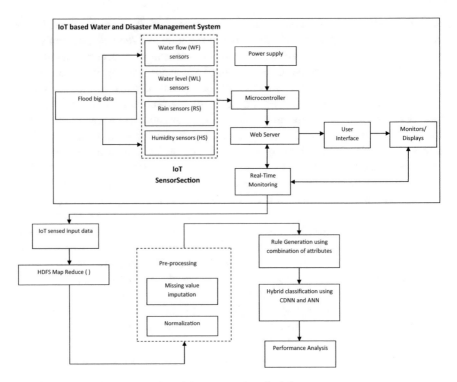

Fig. 11.1 Schematic representation of the suggested methodology

water sensor and a rainfall sensor as well as a humidity sensor are all part of the flood monitoring module. Other components include diodes, a PCB with an LCD and a transformer/adapter as well as buttons, switches, and an IC. In addition to the Raspberry Pi, four sensors will be put at different positions to monitor the data. Water levels in the water body rise, and the buzzer beeps, alerting you to the remaining time until the region floods.

Figure 11.3 shows the suggested IoT circuit prototype's schematic description and schematic diagram. IoT components will be well-connected to the health care system in this scenario. Ultrasonic sensors measure water levels in the pipe, and a rain sensor detects the weather. The Arduino Uno microcontroller, integrated with an Ethernet shield, is used to analyze additional data such as height and rain. Over the web server, the results of data processing will be dispersed the IP address of the web server is utilized to access the flood monitoring information system, which in this case is 192.168.0.4. Using IP address access, you may find out about the rainy circumstances on the web using web flood information provided by IP address access Flood warning system in real-time is developed utilizing HTML and JavaScript programming, related to C language integrated in Arduino Uno's microcontroller board.

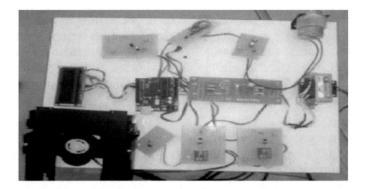

Fig. 11.2 Experimental set up

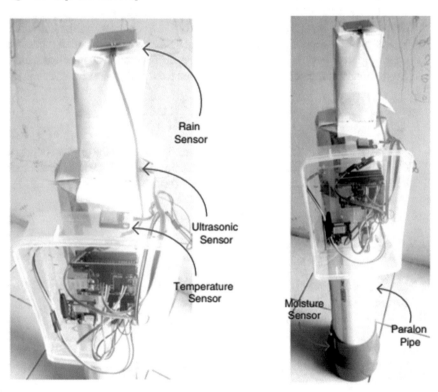

Fig. 11.3 System prototype circuit

Construction of a 5-in. paralon pipe flood height utilizing an ultrasonic sensor and rain sensors on the upper side of the pipe As a cross-section, a cork float is inserted within the ultrasonic sensor trigger component to reflect the echo signal. Buoyancy increases with the height of the floodwaters. Early warning information on floods, including flood information and rain conditions in real-time, as well as risk information.

(a) **Normalization**

This may be done by using the HDFS map reduction framework in conjunction with the data analysis in order to extract the specialized functions needed for categorization. Because it distributes data into small clusters, Map Reduce Framework provides optimistic outcomes and aids in the processing of large data sets. Subgroups are created from the huge volume of data. Interrelationships between the subgroups must be assessed in order to avoid major problems. Group the data according to its size and kind. After clustering, the individual values may be found. Utilizes the patented matrix vectors to determine the matrices and project the data into a new subspace that is equivalent or less scaled. An optimally comparable subset of variables from a big dataset is selected.

$$F = \frac{\lambda_1 + \lambda_2 + \lambda_3 \dots \lambda_k}{\lambda_1 + \lambda_2 + \dots + \lambda_k + \dots \lambda_d} \tag{11.1}$$

λ is the Eigen value; F is the new set of data features; d is the original features. To reflect the data records with low-dimensional vectors, we need to estimate n Eigen vectors that lead to the highest Eigen values.

$$\phi = l_1, l, \dots l_m$$

$$A = \mathrm{diag}(\lambda_1, \lambda_2, \dots \lambda_m) \tag{11.2}$$

Here, A defines the diagonal matrix of error values and we have received the covariance matrix as,

$$K\phi = \phi A \tag{11.3}$$

Then determine the number of features by using the following criterion

$$\frac{\sum_{i=1}^{T} \lambda_i}{\sum_{i=1}^{P} \lambda_i} > M \tag{11.4}$$

P is the total number of error values. The pointed errors gets extracted.

A normalization of the data and missing values is required after eliminating all mistakes and missing values. Prior to computing the average, it is necessary to normalize numbers obtained at different balances to a theoretically similar scale. Only a rescaling procedure is required for specific normalization forms in order to get values associated with certain other elements. Due to the familiarity with data parameters, we can easily correct our blunders. In most cases, the sensor data values will not be randomly distributed once the mistakes have been corrected for. It is necessary to obtain the z-point as a preliminary step in the normalizing procedure. As seen in Eq. 11.5, the Z-score will be expressed.

$$Z = [(G - \mu)/\sigma]$$ (11.5)

where μ is the mean of the data and σ is the standard deviation of the sensor data.

$$Z = \frac{G - \overline{G}}{S}$$ (11.6)

where $\overline{G}$ is the mean of the sample and S is the standard deviation of the sample. Then the Hat matrix (H) can be calculated by using Eq. (11.7)

$$H = L * (G^T G)^{-1 G^T}$$ (11.7)

The variance for the Hat matrix is represented in Eq. 11.8

$$Var(\hat{\epsilon}_i) = \sigma^2 (1 - h_{ii})$$ (11.8)

$$Var(\hat{\epsilon}_i) = \sigma^2 \left(1 - \frac{1}{n} - \left[(x_i - \overline{x}^2) / \sum_{j=1}^{i} (x_j - \overline{x}^2) \right] \right)$$ (11.9)

Then the residual can be calculated by using Eq. (11.10)

$$t_i = \frac{\hat{\epsilon}_i}{\overline{\sigma}} \sqrt{1 - h_{ii}}$$ (11.10)

where $\overline{\sigma}$ is an estimate of the σ which is the mean values

$$o^k = \sqrt{(E(G - \mu)^K)^2}$$ (11.11)

where L is a random variable and E is the expected value.
Use the mean μ particularly for the usually ordered distribution to standardize the distribution of the variable.

$$C_v = \frac{S}{\bar{\bar{G}}} \tag{11.12}$$

where C_v is the coefficient of the variance.

Using the feature scale technique, all values from 0 to 1 may be obtained. As a result of this method, standardization based on continuity is used. The normalized equation will take the form G'

$$G' = \frac{(G - G_{\min})}{(G_{\max} - G_{\min})} \tag{11.13}$$

The spectrum and data variability will be equalized as a result of standardizing the data in the data collection. The redundancy of data has been minimized to a large extent or eliminated completely. They may then be used as input in the compilation process.

(b) **Classification**

The initial stage of the classification was the rule generation for the classifiers.

Rule Generation

Here's a manufacturing method that uses classifier systems to design the rule robotically. The planned servers will reliably check if a file has been received before releasing it. The rule generating procedure is shown in Fig. 11.4. This system analyses the information once it has been received and determines whether or not it is raining. In addition, the rule will produce a new rule, and if the old rule should be stored in the categorization system, it will be compared with the new rule. If the rule already exists, the new rule produced will go unnoticed because the old one already exists. This rule will be entered in the system if it has never been there before. As soon as a package is not banned, it is sent back to the originating server. Upon the entry of a new file, the procedure will be quickly resumed and completed. So that file verification takes less than a minute, the system will have preplanned and scheduled the procedure. Defined as an important step, rule generation may be coded into a module to produce rules automatically based on a file.

Hybrid *CDNN and ANN* is utilized in the flood data classification Framework for accurate classification of the hazardous situation from the dense data by utilizing the features of a sensor data, i.e., $i, j \ldots k$ to train classifier $\pi_{i,j}$ using C as the positive set then compute the distance from each subject to the separating hyperplane of each classifier for each subject there is a vector $d = (d_1, d_2, \ldots, d_k)$, where d_i denotes the distance of the subject to the separating hyper plane of the ith classifiers. For a two-class training example, k solves the equation below. The distance between the data i and j is computed with the tth metric as

$$d_{i,j} = \text{trace}(Mt(x_i - x_j)(x_i - x_j)^T) \tag{11.14}$$

$$A(s) = [0, \text{Csum}(2r(s_1) - 1) \ldots \text{Csum}(2r(s_m) - 1)] \tag{11.15}$$

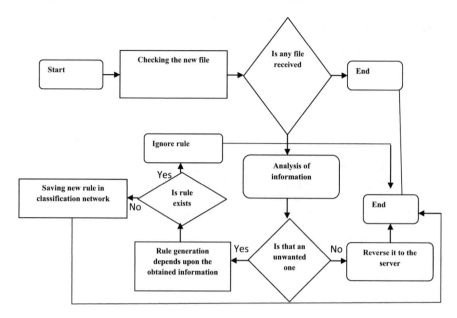

Fig. 11.4 Process of classification rule generation

In which Csum defines the sum in a cumulative way, m is the iteration at the maximum, s denotes random motion steps and $r(s)$ denotes the function in a stochastic manner (11.16):

$$r(s) = \begin{cases} 1 \text{ if } r_{no} > 0.5 \\ 0 \text{ if } r_{no} \leq 0.5 \end{cases} \tag{11.16}$$

In shown in Eq. (11.17), A^l and B^l denote the lowest value and the extreme value of the variable at the lth iteration.

$$z = 10^y \frac{u}{U} \tag{11.17}$$

which z is the ratio of flood level.

In this, u denotes its current iteration and U refers to the hazard.

As soon as the distance can gets calculated classifier realizes the hazardous situation it will classifies the hazardous level, (danger or wary or secure)

$$A^l = \frac{A^l}{Z} \tag{11.18}$$

$$B^l = \frac{B^l}{Z} \tag{11.19}$$

Algorithm 1 (Hybrid **CDNN** and ANN)

Input: details of rain data
Output: Flood prediction

Initialize all weights and prejudice in the Network
while termination is not fulfilled X in $D/$ {
   //for the data proliferation is:
     for each input-layer in unit j { $OJ = Ij$;
     //output of the input unit is the real input value of each unit secret or output-layer input unit J
{
/ /Compute the input net in unit j in relation to the previous input-layer;
For the output layer of each unit $r(s) = -\emptyset(\|W\|)^2$
   //Calculate the errors of each unit j of the hidden layers from last up to first hidden layer—
   //Calculate an flood level of each unit j of the hidden layers,
$z = 10^y \frac{u}{U} Z$ of each network weight $r(s)$.
// Calculate the hazardous level in the next layer/
}}
End
End

Result and Discussion

It also has a moving height-of-the-water animation, as seen in Fig. 11.5, where the flood height is 32 cm and the danger level is displayed, as well as information on the wet conditions.

Data can be analyzed by the classifier based on their input. Figure 11.5 shows how the flood categorization details were determined from the data. On the basis of a comparison of the performance of the proposed classification technique to other current methods on data classification [16]. True Positive (TP) is the condition that the quantity and accuracy of instances are categorized. False Positive (FP) is the circumstance in which the number of cases that were really wrong is categorized as exact. The condition is that False Negative (FN) classifies the number of instances as false while true. True Negative (TN) is a prerequisite for classifying the amount of records as wrong when they were untrue.

As of from Fig. 11.6 the positive prediction of the fold condition was higher in the proposed methodology than any other existing methodologies.

False positive is a binary classification mistake in which a test result wrongly indicated the existence of a flood condition if the flood was not happen. Here from Fig. 11.7 the suggested methodology have very less false positive rate than any other existing methodologies.

The comparative accuracy analyses for both the proposed and current approaches are shown in Fig. 11.8. The x-axis shows the data volume, while the y-axis shows the exactness. The result shows that the mechanism presented can provide greater classification accuracy than the conventional techniques.

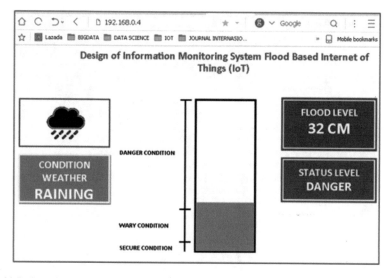

Fig. 11.5 Overall output

Fig. 11.6 Data versus TP rate

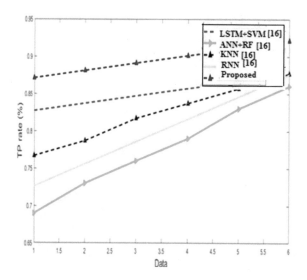

Figure 11.9 showing that the proposed methodology express maximum precision yield of 91.2%, which is better than other methods.

Figure 11.10 showing that the proposed methodology express maximum recall yield of 91.3%, which is better than other methods. When compared to other existing methodology the suggested methodology outperformed well.

Fig. 11.7 Data versus FP
rate

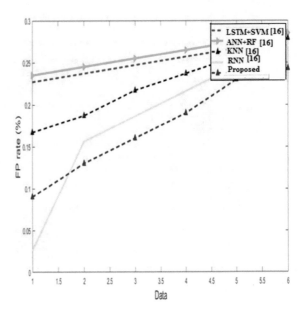

Fig. 11.8 Accuracy
comparisons with proposed
work

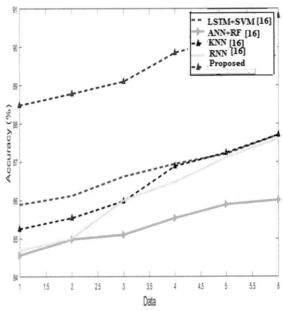

Fig. 11.9 Data versus precision

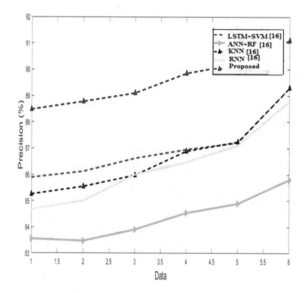

Fig. 11.10 Data versus recall

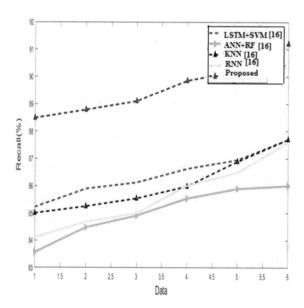

Conclusion

An integrated data extraction tool on a web-based system may be used to investigate flood monitoring and early warning systems for particular research initiatives, such as the Sentinel Asia Project. As a result, crisis management will be more effective than ever before. In addition, the availability of compromised data and the growth of the local area network are crucial. Weather and hydrological information from in-situ data will be included into future developments, allowing for a more accurate flood analysis. A more accurate, trustworthy, and efficient forecasting system will be created. The gap in the disaster management system, however, can be filled by a variety of variables. Water runoff and river levels, for example, will need to be considered. Other than that, the local organization should automatically receive and incorporate the sensor data and supporting data (such as land use/land cover, road) as part of the data needed for an evacuation and management plan. Using a web-based system to evaluate and extract information is the research's uniqueness and core. Goals include creating an easy to use, low-cost, data processing interface that can identify flooded regions at the earliest possible time. Technical and non-technical users alike can benefit from the suggested solution. There are a variety of ways to use the system idea. Its idea will extend the research study field for the future development of real-time disaster monitoring systems, according to the authors. There is a shift in the system from manual detection to automated detection and delivery. As a realistic plan, it is predicted to be successful. A wide range of operating systems were supported by this system, which was meant to provide a rapid processing service and an interesting user experience. The proposed study proposes a Hybrid CDNN-ANN classification technique for managing big dataset samples. As a result of the use of a clustering approach paired with a hybrid CDNN-ANN classifier, a trained persistent classifier can reliably predict crime using a rainfall dataset. The followings are issues that need to be further discussed in the future.

- The Sensor Web Enablement (SWE): The Open Geospatial Consortium, Inc. started the group (OCG). How to develop standard standards for connecting earth observation equipment with web-based will be the focus of the research group Efforts to obtain satellite pictures using sensor observation systems, using OGC measurement standards, will be closely followed in the coming months. For the user who has no photographs in hand, the objective is accessing sensor data.
- Hydrological Data: The quantity of water being discharged into the river will be calculated based on the water level and rainfall data obtained from the field. A final evaluation and integration of the entire quantity released will validate the information collected from sensors.
- Areas Extraction Algorithms: At order to get better results, the inundated regions extraction algorithm has to be improved in high altitude areas. Various satellite pictures, such as Aster images, should be examined.
- Active Communication and Publication: Authorities and local residents must work closely together to develop a comprehensive strategy that will produce timely warnings. Provide users with practical and useful information and encourage them to use the system.

References

1. Mendoza-Cano, O., Aquino-Santos, R., López-De la Cruz, J., Edwards, R.M., Khouakhi, A., Pattison, I., et al.: Experiments of an IoT-based wireless sensor network for flood monitoring in Colima, Mexico. J. Hydroinf. **23**, 385–401 (2021)
2. Rani, D.S., Jayalakshmi, G., Baligar, V.P.: Low cost IoT based flood monitoring system using machine learning and neural networks: flood alerting and rainfall prediction. In: 2020 2nd International Conference on Innovative Mechanisms for Industry Applications (ICIMIA), pp. 261–267 (2020)
3. Shah, W.M., Arif, F., Shahrin, A., Hassan, A.: The implementation of an IoT-based flood alert system. Int. J. Adv. Comput. Sci. Appl. **9**, 620–623 (2018)
4. Satria, D., Yana, S., Munadi, R., Syahreza, S.: Prototype of Google maps-based flood monitoring system using Arduino and GSM module. Int. Res. J. Eng. Technol **4**, 1044–1047 (2017)
5. Ragnoli, M., Barile, G., Leoni, A., Ferri, G., Stornelli, V.: An autonomous low-power lora-based flood-monitoring system. J. Low Power Electron. Appl. **10**, 15 (2020)
6. Sood, S.K., Sandhu, R., Singla, K., Chang, V.: IoT, big data and HPC based smart flood management framework. Sustain. Comput. Inform. Syst. **20**, 102–117 (2018)
7. Chaduvula, K., Markapudi, B.R., Jyothi, C.R.: Design and Implementation of IoT based flood alert monitoring system using microcontroller 8051. Mater. Today: Proc. (2021)
8. Mousavi, F.S., Yousefi, S., Abghari, H., Ghasemzadeh, A.:Design of an IoT-based flood early detection system using machine learning. In: 2021 26th International Computer Conference, Computer Society of Iran (CSICC), pp. 1–7 (2021)
9. Kumar, M.A., Singh, V., Chaubey, V.K., Singh, V.: Iot Based Flood Detector and Early Precaution Method Instructor to Avoid Any Hazardous Event
10. Khalid, M.S.B., Shafiai, S.B.: Flood disaster management in Malaysia: An evaluation of the effectiveness flood delivery system. Int. J. Soc. Sci. Humanity **5**, 398 (2015)
11. Bande, S., Shete, V.V.:Smart flood disaster prediction system using IoT & neural networks. In: 2017 International Conference on Smart Technologies for Smart Nation (SmartTechCon), pp. 189–194 (2017)
12. Wijekoon, N.S., Pallewatte, C.D., Sandareka, P., Pradeep, R.: Integrated Solution System for Flood Disaster Management (using IoT and GIS)
13. Atmojo, P.S., Sachro, S.S.: Disaster management: selections of evacuation routes due to flood disaster. Proc. Eng. **171**, 1478–1485 (2017)
14. Ghapar, A.A., Yussof, S., Bakar, A.A.: Internet of Things (IoT) architecture for flood data management. Int. J. Future Gener. Commun. Netw **11**, 55–62 (2018)
15. Sumitra, I., Supatmi, S.: Mamdani fuzzy inference system using three parameters for flood disaster forecasting in Bandung region. In: IOP Conference Series: Materials Science and Engineering, p. 042008 (2019)
16. Khalaf, M., Alaskar, H., Hussain, A.J., Baker, T., Maamar, Z., Buyya, R., et al.: IoT-enabled flood severity prediction via ensemble machine learning models. IEEE Access **8**, 70375–70386 (2020)

Chapter 12
ANN: Concept and Application in Brain Tumor Segmentation

Amit Verma

Introduction

Nowadays, artificial network (AI) has become a part of almost all the fields to improve accuracy and to automate the work, with higher precision. In the medical field, still there are lot of areas in which AI can play a vital role in better diagnosis. Brain tumor segmentation is one of the emerging areas in which many researchers are doing work to automate the process of brain tumor segmentation based on Medical Resonance Imaging (MRIs). In a broad sense, the term brain tumor segmentation means identifying the presence or absence of tumor in MR image. It is not restricted only to identifying the presence or absence of the tumor but also AI-based algorithm can identify the exact area of the tumor that can help doctors for better diagnosis. Until today, various automatic and semi-automatic methods proposed for the segmentation process [1–3]. The fully automatic [4, 5] approach requires a very high computation time and the seed pixels are automatically selected by the algorithm. Whereas in the semi-automatic method [6–8] users manually select the Region of Interest (RoI) to provide as input for the execution of the algorithm, therefore, the computational time is lesser as compared to the fully automatic approach. Artificial Neural Network (ANN) is widely used by many researchers to automate/semi-automate the process of brain tumor segmentation with higher accuracies. ANN is based on the working and structure of the mind as shown in Fig. 12.1, where the basic structure of the brain resembles the neural network. The dendrites act as input layer in ANN that takes the input from the user and provides it to the neuron, which processes it and sends it to the output layer. In shown in Fig. 12.1b, there are three inputs, given to the neuron for some processing and have some specific weights. Neuron process the information

A. Verma (✉)
School of Computer Science, UPES, Dehradun, Uttarakhand, India
e-mail: amit.verma@ddn.upes.ac.in

© The Author(s), under exclusive license to Springer Nature Singapore Pte Ltd. 2023 175
V. Kadyan et al. (eds.), *Deep Learning Technologies for the Sustainable Development Goals*, Advanced Technologies and Societal Change,
https://doi.org/10.1007/978-981-19-5723-9_12

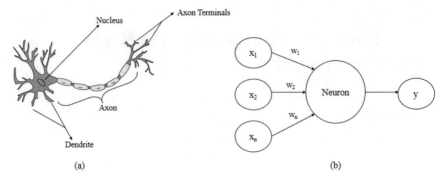

Fig. 12.1 Brain neuron structure depicted as ANN basic structure

and provide the result. Figure 12.1b represents a structure of a neural network with no hidden layer, also called perceptron (or single-layer neural network) [9–12].

ANN used in extracting information from medical images for segmenting and classifying of brain tumors in MRIs. ANN-based methods produce better results as compared to machine learning methods, mainly two main methods are used for the segmentation/classification approach that is patch based and semantic segmentation [13–17]

Concept of ANN and Activation Function

A very basic or single-layer neural network comprises the input layer, neuron, and output layer. Neuron process the information given by the input layer on the basis of assigned weights and provide the output y'. Every neuron performs two steps first as a summation of input values and weights and applies the activation function on the summation as shown in Fig. 12.2. In step 1, the summation of input values and weights are performed as shown in Eq. 12.1 represented by S where n is a number of features $(x1, x2, ..., xn)$.

Fig. 12.2 Neuron

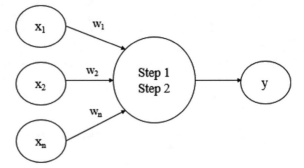

$$S = \sum_{i=1}^{n} W_i X_i \qquad (12.1)$$

Step 2 is an application of the activation function to activate the particular neuron. The activation function decides whether the neuron will be activated or remain deactivated based on the value of S. During the forward propagation from left to right, the activation function decided which of the following neurons will take part in predicting the final result y' as shown in Fig. 12.3. Now, y' (predicted result) is compared with the actual y for that input (single row of the dataset). Now the weights are adjusted to minimize the difference between y' and y which is represented by C as shown in Eq. 12.2. The equation is squared to remove the negation sign. And, the process of adjusting weight continues till the value of C become almost zero.

$$C = \frac{1}{2}(y' - y)^2 \qquad (12.2)$$

This process of adjusting the weights recursively to minimize C that is Mean Squared Error (MSE) is called backpropagation [18, 19].

Fig. 12.3 Process of forward propagation based on activation function

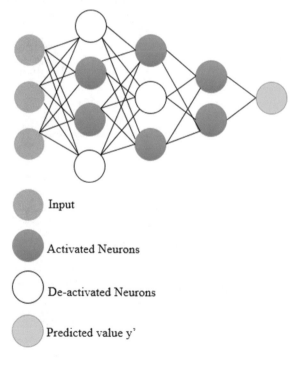

Input

Activated Neurons

De-activated Neurons

Predicted value y'

Activation Function (Φ)

Based on activation function it is decided whether the particular neuron will contribute to the prediction of the result or not. Activation function φ is applied on the value of S. There are many activation functions that can be used according to the problem [20–25]

Threshold Function

It is used for binary classification problem that is in which the result either of the two classes (eg. 0 or 1, A or B). The threshold function cannot be used with multi-class problems. The graph of threshold function is shown below in Fig. 12.4, in which is the y-axis is representing the predicted output y' and the x-axis is representing the summation of input and assign weights. Now, let us take the problem of finding the correct output of AND operation using neural network and threshold function. Considering Table 12.1 of AND operation as now we will design a neural network to predict the result on the basis of the input and weights.

Considering Fig. 12.5, the first node is the biased unit and the other two nodes are input and weights are assigned with blue color that is 10 to each connection. Now, when the first row is given as input to the network then in neuron the S will be calculated and the threshold activation function is applied to it to get the value of y'.

$$S = 10 + 10.x1 + 10.x2$$
$$S = 10 + 10.0 + 10.0$$

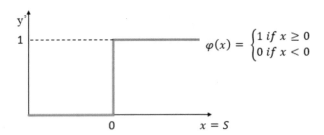

$$\varphi(x) = \begin{cases} 1 \ if \ x \geq 0 \\ 0 \ if \ x < 0 \end{cases}$$

Fig. 12.4 Threshold activation function

Table 12.1 AND operation

X1	X2	Y
0	0	0
0	1	0
1	0	0
1	1	1

Fig. 12.5 Neural network
for the first input

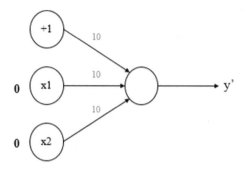

$$S = 10$$

Now, the activation function is applied on the value of S

$$\varphi(S) = \varphi(10)$$

According to the Fig. 12.4, the value of $\varphi(10)$ will be equal to 1, so y' for this particular input will be 1. But, according to Table 12.1 it must be equal to 0.

Therefore, according to Eq. 12.2 the value of $C = 1/2$ which is very high and now to minimize the value of C backpropagation is done to adjust the weights as shown in Fig. 12.6.

Now, again we calculate the new y' with new weights as shown in below lines

$$S = -30 + 20.x1 + 20.x2$$
$$S = -30 + 20.0 + 20.0$$
$$S = -30$$

Now, the activation function is applied on the value of S

$$\varphi(S) = \varphi(-30)$$

Fig. 12.6 Neural network
with new weights

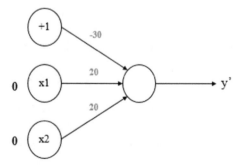

Fig. 12.7 Second input is
given in input layer

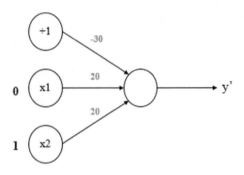

According to Fig. 12.4, the value of $\varphi(-30)$ will be equal to 0, so y' for this particular input will be 0. And as the actual value of y for this particular input is also 0; therefore, we will consider these weights and calculate the value for another set of inputs (second row) as shown in Fig. 12.7.

Again the value of S and $\varphi(S)$ will be calculated to compute the value of y'.

$$S = -30 + 20.x1 + 20.x2$$
$$S = -30 + 20.0 + 20.1$$
$$S = -10$$

Now, the activation function is applied on the value of S

$$\varphi(S) = \varphi(-10)$$

According to Fig. 12.4, the value of $\varphi(-10)$ will be equal to 0, so y' for this particular input will be 0. And as the actual value of y for this particular input is also 0; therefore, we will consider these weights and calculate the value for another set of inputs (third row).

$$S = -30 + 20.x1 + 20.x2$$
$$S = -30 + 20.1 + 20.0$$
$$S = -10$$

Now, the activation function is applied on the value of S

$$\varphi(S) = \varphi(-10)$$

Again the value of $\varphi(-10)$ is 0, so y' for this particular input will be 0. And as the actual value of y for this particular input is also 0; therefore, we will consider these weights and calculate the value for another set of inputs (final row).

$$S = -30 + 20.x1 + 20.x2$$

$$S = -30 + 20.1 + 20.1$$
$$S = 10$$

Now, the activation function is applied on the value of S

$$\varphi(S) = \varphi(10)$$

According to Fig. 12.4, the value of $\varphi(10)$ will be equal to 1, so y' for this particular input will be 1. And as the actual value of y for this particular input is also 1; therefore, our model is working correctly and accurately for predicting the result of AND operation.

Sigmoid Function

The working of sigmoid function is same as discussed in the previous section, the only difference is the graph or representation of sigmoid function is different as shown in Fig. 12.8. According to the figure, y-axis is representing the predicted output y' and x-axis is representing the summation of input and assign weights.

In this case, we will see the working of sigmoid function for the same AND logic and consider the weighted neural network as shown Fig. 12.7. When the first row of Table 12.1 is given as input to the neural network than in the neuron the value of S and $\varphi(S)$ will be get calculated.

$$S = -30 + 20.x1 + 20.x2$$
$$S = -30 + 20.0 + 20.0$$
$$S = -30$$

According to the sigmoid graph as shown in Fig. 12.8 as the value of $S = -30$ that is $x = -30$ as the value x is far left from 0; therefore, the value of $\varphi(x)$ would be 0 that is $y' = 0$. And as the actual label that is the value of y for the particular row is also 0; therefore, the model is working great with these weights. Similarly, the second row is given as input to the neural network and the value of S and $\varphi(S)$

Fig. 12.8 Representation of sigmoid activation function

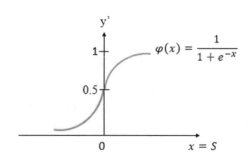

will be get calculated.

$$S = -30 + 20.x1 + 20.x2$$
$$S = -30 + 20.0 + 20.1$$
$$S = -10$$

According to the sigmoid graph still the value x is at the left from 0; therefore, the value of $\varphi(x)$ would be near to 0 that is $y' = 0$. Following the similar process, the output would be predicted for third input row

$$S = -30 + 20.x1 + 20.x2$$
$$S = -30 + 20.1 + 20.0$$
$$S = -10$$

Again the value of x is at the left from 0; therefore, $\varphi(x)$ would be near to 0 that is $y' = 0$. Finally, we calculate for the last input row.

$$S = -30 + 20.x1 + 20.x2$$
$$S = -30 + 20.1 + 20.1$$
$$S = 10$$

In this case, the value of $S = 10$ and the value of $x = 10$ which would be right of the 0 according to the sigmoid graph. Therefore, the value of $\varphi(x)$ would be near to 1 that is $y' = 1$.

Hyperbolic Tangent (tanh)

This activation function is also very similar to the sigmoid function, the only difference is it takes any input value and scale it with in the range of -1 to 1, whereas sigmoid function scale the input in the range of 0 to 1. If the input is a greater real number than the output would be close to 1 and if the input to the activation function is greater negative number than the output of the activation function will be close to -1 as shown in Fig. 12.9.

Rectifier Function (ReLU)

It just propagate the positive input that is if the value of S is greater than 0, where S is the summation of product of weights and inputs as shown in Eq. 12.1. The graph of ReLU is shown in Fig. 12.10 which show that if the value of $x > 0$ than the value of $\varphi(x)$ would be equal to x, where $x = S$.

Fig. 12.9 Representation of ReLU activation function

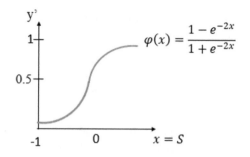

$$\varphi(x) = \frac{1 - e^{-2x}}{1 + e^{-2x}}$$

Fig. 12.10 Representation of rectifier activation function

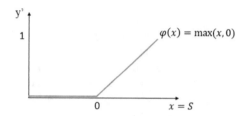

$$\varphi(x) = \max(x, 0)$$

ReLU activation function is one of the most popular function used in neural network with multiple layers or we can say it is a default activation function in neural network with multiple layers. As threshold activation function can only be used for binary classification problem, sigmoid and tanh activation function face the problem of vanishing gradient [26] during back propagation for adjusting the weights.

Steps Involved in ANN

Here, we will discuss the basic steps required for the application of ANN for any particular problem.

Step 1—Randomly assign the weights to the first layer, weights are recommended to be close to 0 but not 0 it could be in range from −0.5 to +0.5.

Step 2—Input the first row of the dataset to the input layer and the number of nodes in the input layer should be equal to the number of independent input that is number of features or number of columns.

Step 3—Now, propagate the network from left to right by assigning weights and applying the activation function until we get y' that is the predicted output. This process is called forward propagation.

Step 4—Now, calculate the difference of predicted y' with the actual y and adjust the weights from last layer to the first layer more generally from right to left. The adjustment of weights should be done in such a way that error get minimized. This process of adjusting the weights from right to left is called back propagation.

Step 5—Now, repeat above 4 steps, it can be done either for every input at once and adjust the weights after that or the batch of inputs at once and update the weights after getting the error of particular batch.

Step 6—After one epoch, the process can be repeated for multiple epoch until we get the global minima. Here, epoch means that we get the neural network (with weights) after inputting all the data in the network. More epochs can be done until we reach the global minima.

Application of ANN in Brain Tumor Segmentation—Image segmentation defined as a process of partitioning segmenting the area of interest from the image. It is having a wide range of applications in the field of medical science. Majorly, used in brain tumor segmentation, tissue classification, fracture detection, tumor volume calculation, lung tumor segmentation and many more. A brain tumor considered as an abnormal growth of tissues in any random shape. It can be malignant or benign, only malignant cells are cancerous that multiplied or grow at a very high rate. MRI considered as an evolutionary development in the field of brain tumor segmentation. MRIs provide significant information about the brain tissues with different modalities like FLAIR, T1, T1c, and T2. Various modalities with segmented mask of MR image are shown in Fig. 12.11 as per BraTS 2013 dataset [27]. These MRI are analysis by the radiologist for manually or knowledge base segmentation of brain tumor to generate the report for the same. These reports are very much to the human error. Moreover, manual segmentation of brain tumor based on MR images is very tedious task. As these reports play a pivotal role in the diagnosis of the patient. Therefore, the accuracy of brain tumor segmentation and uniformity of reports from different MRI machine are two major issues in the field of medical science.

Considering the above requirement, it is indispensable to have an automated approach for segmenting the brain tumor MR images with high accuracy. Therefore, many researchers has contributed various semi-automated and fully automated methods for brain tumor segmentation. With the evolution of machine leaning, many researchers propose various methods with the remarkable results in brain tumor segmentation. Support vector machine, random forest, KNN, and K-mean clustering are some majorly used machine learning algorithms used for brain tumor segmentation. However, the accuracy of machine learning based models have great scope

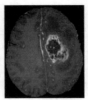

Fig. 12.11 Example of brain tumor MRI modalities FLAIR, T1, T1c, and T2

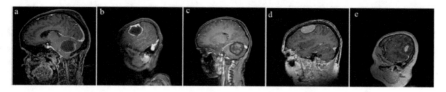

Fig. 12.12 Brain tumor segmentation

of improvement. Moreover, even after providing large amount of data the accuracy of machine learning models remain stagnant. With the profuse of high quality MRI data availability, neural network become emerging approach for various researchers.

Artificial neural network, adopted as promising approach by the researchers to proposed automatic method for brain tumor segmentation. In addition, used for pixel wise binary classification that is tumor or non-tumor [28–30]. Moreover, multi-class classification models used ANN results in higher accuracy as compare to various outperforming machine learning methods. ANN is a complex network of artificial neurons and multiple hidden layers. The input given to the input layer, traverse through the web of hidden layers before giving a relevant output. Forward and backward propagation used to fine tune the network for better accuracy or minimize loss. In [31], the author has proposed a novel method for brain tumor segmentation, by fusing SVM and ANN. The segmented brain tumor in MR images shown in Fig. 12.12.

Feed-forward neural network (FNN) is one of the simple ANN used for brain tumor segmentation [32–35]. Information flow from input to output layers, traversing through multiple hidden layers. FNN model can be trained using forward and backward propagation approach. Forward neural network [36] proven better than some standard machine learning algorithms like Bayesian and KNN with accuracy of 80%. In [37], author proposed a brain tumor segmentation method based on feedback pulse coupled neural network, features of the images extracted using DWT. Moreover, dimensionality also reduced with principle component analysis, to improve the computation process. Shenbagarajan et al. [38] proposed a novel model for brain tumor segmentation by integrating region-based active contour model for segmentation and ANN for classification.

Conclusion

In this, we discussed the concept and importance of artificial neural networks in solving complex problems. The role and types of activation function are discussed in detail, and which activation function should be used for a particular problem. Artificial neural network plays a vital role in automating the process of segmenting various medical images, for example, brain tumor segmentation. With this automation, uniformity can be maintained in the results of MR images so that doctors can diagnose the patient in a much better way. Activation function is discussed, for example, as the activation function plays a major role in designing any neural network. The concept of front and backpropagation is also discussed. In the future, deep learning and machine learning algorithms can be used in various other medical areas which are still unexplored by many researchers to give some useful contribution in medical field.

References

1. Clark, M.C., Hall, L.O., Goldgof, D.B., Velthuizen, R., Murtagh, F.R., Silbiger, M.S.: Automatic tumor segmentation using knowledge based techniques. IEEE Trans. Med. Imaging **17**(2), 187–2011 (1998)
2. Lynn, M., Lawerence, O.H., Demitry, B., Goldgof, R.M.F.: Automatic segmentation of non-enhancing brain tumors. Artif. Intell. Med. **21**, 43–63 (2001)
3. Wang, T., Cheng, I., Basu, A.: Fluid vector flow and applications in brain tumor segmentation. IEEE Trans. Biomed. Eng. **56**(3), 781–789 (2009)
4. Salah, M.B., et al.: Fully automated brain tumor segmentation using two mri modalities. In: International Symposium on Visual Computing. Springer, Berlin (2013)
5. Alqazzaz, S., et al.: Automated brain tumor segmentation on multi-modal MR image using SegNet. Computat. Visual Media **5**(2), 209–219 (2019)
6. Dubey, R.B., et al.: Semi-automatic segmentation of MRI brain tumor. ICGST-GVIP J. **9**(4), 33–40 (2009)
7. Lim, K.Y., Mandava, R.: A multi-phase semi-automatic approach for multisequence brain tumor image segmentation. Expert Syst. Appl. **112**, 288–300 (2018)
8. Guo, X., Schwartz, L., Zhao, B.: Semi-automatic segmentation of multimodal brain tumor using active contours. Multimodal Brain Tumor Segmentation **27** (2013)
9. Raudys, Š: Evolution and generalization of a single neurone: I. Single-layer perceptron as seven statistical classifiers. Neural Netw. **11**(2), 283–296 (1998)
10. Shynk, J.J.: Performance surfaces of a single-layer perceptron. IEEE Trans. Neural Netw. **1**(3), 268–274 (1990)
11. Auer, P., Burgsteiner, H., Maass, W.: A learning rule for very simple universal approximators consisting of a single layer of perceptrons. Neural Netw. **21**(5), 786–795 (2008)
12. Marcialis, G.L., Roli, F.: Fusion of multiple fingerprint matchers by single-layer perceptron with class-separation loss function. Pattern Recogn. Lett. **26**(12), 1830–1839 (2005)
13. Kao, P.-Y., et al.: Improving patch-based convolutional neural networks for MRI brain tumor segmentation by leveraging location information. Front. Neurosci. **13**, 1449 (2020)
14. Cordier, N., et al.: Patch-based segmentation of brain tissues. MICCAI challenge on multimodal brain tumor segmentation. IEEE (2013)
15. Sharif, M., et al.: A unified patch based method for brain tumor detection using features fusion. Cogn. Syst. Res. **59**, 273–286 (2020)
16. Rezaei, M., et al.: A conditional adversarial network for semantic segmentation of brain tumor. In: International MICCAI Brainlesion Workshop. Springer, Cham (2017)
17. Pereira, S., Alves, V., Silva, C.A.: Adaptive feature recombination and recalibration for semantic segmentation: application to brain tumor segmentation in MRI. In: International Conference on Medical Image Computing and Computer-Assisted Intervention. Springer, Cham (2018)
18. Hecht-Nielsen, R.: Theory of the backpropagation neural network. Neural Networks for Perception, pp. 65–93. Academic Press (1992)
19. Li, J., et al.: Brief introduction of back propagation (BP) neural network algorithm and its improvement. In: Advances in Computer Science and Information Engineering, pp. 553–558. Springer, Berlin (2012)

20. Plagianakos, V.P., Vrahatis, M.N.: Training neural networks with threshold activation functions and constrained integer weights. In: Proceedings of the IEEE-INNS-ENNS International Joint Conference on Neural Networks. IJCNN 2000. Neural Computing: New Challenges and Perspectives for the New Millennium, vol. 5. IEEE (2000)
21. Huang, G.-B., et al.: Can threshold networks be trained directly? IEEE Trans. Circuits Syst. II Express Briefs **53**(3), 187–191
22. Jamel, T.M., Khammas, B.M.: Implementation of a sigmoid activation function for neural network using FPGA. In: 13th Scientific Conference of Al-Ma'moon University College, vol. 13 (2012)
23. Pratiwi, H., et al.: Sigmoid activation function in selecting the best model of artificial neural networks. J. Phys. Conf. Series **1471**(1) (2020)
24. Sartin, M.A., Da Silva, A.C.R.: Approximation of hyperbolic tangent activation function using hybrid methods. In: 2013 8th International Workshop on Reconfigurable and Communication-Centric Systems-on-Chip (ReCoSoC). IEEE (2013)
25. Agarap, A.F.: Deep learning using rectified linear units (relu). arXiv preprint arXiv:1803. 08375 (2018)
26. Roodschild, M., Sardiñas, J.G., Will, A.: A new approach for the vanishing gradient problem on sigmoid activation. Prog. Artif. Intell. **9**(4), 351–360 (2020)
27. Hu, K., et al.: Brain tumor segmentation using multi-cascaded convolutional neural networks and conditional random field. IEEE Access **7**, 92615–92629 (2019)
28. Kharrat, A., et al.: A hybrid approach for automatic classification of brain MRI using genetic algorithm and support vector machine. Leonardo J. Sci. **17**(1), 71–82 (2010)
29. Zacharaki, E.I., et al.: Classification of brain tumor type and grade using MRI texture and shape in a machine learning scheme. Magn. Reson. Med. Official J. Int. Soc. Magn. Reson. Med. **62**(6), 1609–1618 (2009)
30. El-Dahshan, E.-S., Hosny, T., Salem, A.-B.: Hybrid intelligent techniques for MRI brain images classification. Digit. Signal Process. **20**(2), 433–441 (2010)
31. Chithambaram, T., Perumal, K.: Brain tumor segmentation using genetic algorithm and ANN techniques. In: 2017 IEEE International Conference on Power, Control, Signals and Instrumentation Engineering (ICPCSI). IEEE (2017)
32. Kavitha, A.R., Chellamuthu, C., Rupa, K.: An efficient approach for brain tumour detection based on modified region growing and neural network in MRI images. In: Int. Conf. Comput. Electron. Electr. Technol. (ICCEET) 1087–1095 (2012)
33. Damodharan, S., Raghavan, D.: Combining tissue segmentation and neural network for brain tumor detection. Int. Arab J. Inform. Technol. **12**(1), 42–52 (2015)
34. Wang, S., Zhang, Y., Dong, Z., Du, S., Ji, G., Yan, J., Yang, J., Wang, Q., Feng, C., Phillips, P.: Feed-forward neural network optimized by hybridization of PSO and ABC for abnormal brain detection. Int. J. Imag. Syst. Technol. **25**(2), 153–164 (2015)
35. Muhammad, N., Fazli, W., Sajid, A.K.: A simple and intelligent approach for brain MRI classification. J. Intell. Fuzzy Syst. **28**(3), 1127–1135 (2015)
36. El-Dahshan, El-Sayed, A., et al.: Computer-aided diagnosis of human brain tumor through MRI: a survey and a new algorithm. Expert Syst. Appl. **41**(11), 5526–5545 (2014)
37. Damodharan, S., Raghavan, D.: Combining tissue segmentation and neural network for brain tumor detection. Int. Arab J. Inform. Technol. (IAJIT) **12**(1) (2015)
38. Shenbagarajan, A., Ramalingam, V., Balasubramanian, C., Palanivel, S.: Tumor diagnosis in MRI brain image using ACM segmentation and ANN-LM classification techniques. Indian J. Sci. Technol. **9**(1)

Chapter 13
Automation of Brain Tumor Segmentation Using Deep Learning

Amit Verma

Abbreviations

BT   Brain tumor
BTS   Brain tumor segmentation
CAD   Computer aided diagnosis
CNN   Convolutional neural network
FCL   Fully connected layer
GAP   Global average pooling
GMP   Global max pooling
ML   Machine learning
MRI   Medical resonance imaging
SOP   Sum of product

Introduction

Nowaday's cancer is one of the widely spreading diseases due to unhealthy food habits, increasing stress in the competitive world, or some other reason. As per the data from ASCO [1] the death rate due to cancer is increasing rapidly. A brain tumor considered as a mass formation due to the abnormal growth of tissues. BT can be broadly classified into two types malignant and benign, in which benign is considered as non-cancerous [2]. Benign BT has very slow or no growth, these tumors can be easily eradicated from surgery and hardly grow again. However, malignant tumors grow very rapidly and contain cancerous tissues, malignant BTs are life threatening

A. Verma (✉)
School of Computer Science, UPES, Dehradun, Uttarakhand, India
e-mail: amit.verma@ddn.upes.ac.in

© The Author(s), under exclusive license to Springer Nature Singapore Pte Ltd. 2023   189
V. Kadyan et al. (eds.), *Deep Learning Technologies for the Sustainable Development Goals*, Advanced Technologies and Societal Change,
https://doi.org/10.1007/978-981-19-5723-9_13

and their growth should be closely monitored for detection of malignant BT in the early stage. Moreover, benign tumors are mostly homogeneous in shape whereas malignant tumors have an irregular share [3]. MRI is one of the successful and commonly used techniques for the detection of BTs [4]. It provides the anatomy of the brain and plays an important role in the analysis of the presence of tumors in the brain. So that the radiologist can see the MRI and segment, the tumor by using some graphical tool like CAD. And the report can be used by the doctors for primary investigation of the patient. The process of identifying and marking the tumor within the MRI and providing its size is called BT segmentation. Which is manually done by the radiologist based on his/her experience [5]. To maintain the uniformity of the report and higher accuracy it is very much required to develop an algorithm for automating the process of BT segmentation. Many researchers have contributed various state-of-art methods to semi-automate and fully automate the process of segmenting tumors using brain MRIs.

Automatic BT segmentation is a challenging task, as it is difficult to differentiate between the malignant and the neighboring healthy tissues due to texture similarity. So, it becomes a fast-growing research area for worldwide researchers to segment the BT with higher accuracy [6]. Various kinds of methods were developed based on machine learning (ML) such as the region growing method which requires selecting the seed points manually [7], level-set method, c-mean, and k-mean clustering-based fuzzy clustering method. Fuzzy-clustering methods also require the knowledge of data [8, 9]. The conventional methods of machine learning require prior knowledge of data distribution and features, so it difficult to build an algorithm to automate the process of brain tumor segmentation. Therefore, deep learning attracts various researchers for this purpose. There are multiple methods developed using deep learning for automating the process of BT segmentation such as generative adversarial networks [10] deep Boltzmann machines, Variational Autoencoder, stacked auto-encoders, CNN, where CNN is the most used algorithm of deep learning in the medical field for automating the process of segmenting the brain tumor In MR images [11]. As compared to machine learning, deep learning can handle much larger data sets and can provide the algorithm for automation extraction of features with minimum human intervene [12]. In the past few decades, multiple deep learning concept CNN [13, 14] based architectures has been developed to automate the process of BT segmentation for better accuracy using the BRATS dataset. In this chapter, we discuss about the introduction to convolutional neural network. Discussed about various layers of CNN like convolution, max pooling, fully connected network, and introduce the application of CNN in brain tumor segmentation.

Convolutional Neural Network

CNN works on tensors [15], tensors can be defined as higher-order matrices such that $y \in \mathbb{R}^{A \times B \times C}$ is tensor of order three where a bold letter is representing the vector symbol, A, B, C are three elements. Every color image in RGB format is order three tensor, previously the color images were converted into 2D images, which cause the loss of color. The loss of color can be a loss of important information especially in medical imaging, this problem of converting the color image to gray scale image solved by using tensors. Input parameters of CNN are tensors, tensors with higher orders are also used in CNN. Tensor can be vectorized [16] in single column as shown below the vectorization of 2×2 matrix. To vectorized 2×2 matrix, first column is vertorized first than second column as shown in Eq. 13.1.

$$A = \begin{bmatrix} 1 & 2 \\ 4 & 5 \end{bmatrix}, vec(A) = \begin{pmatrix} 1 & 4 & 2 & 5 \end{pmatrix}^{\mathrm{T}} \tag{13.1}$$

In general for vectorization of *n-order* tensor, first column is vectorize first than second, and so on after that all vectors are concatenated. CNN learning process use chain rule, suppose $a \in \mathbb{R}$ and $b \in \mathbb{R}$, where 'a' is a scalar unit and 'b' is a vector. Now, let us say 'c' is a function of b, Eq. 13.2 represent the partial derivative of 'c' w.r.t. 'b'.

$$\left| \frac{\partial c}{\partial b_i} \right| = \frac{\partial c}{\partial b_i} \tag{13.2}$$

Now, if any vector is a function of another vector that is $b \in \mathbb{R}^H$ and $d \in \mathbb{R}^W$ and 'b' is the function of 'd'. Than the partial derivative would be represented as shown in Eq. 13.3.

$$\left| \frac{\partial b}{\partial d_j} \right| = \frac{\partial b_i}{\partial d_i} \tag{13.3}$$

Partial derivative of two vectors would be a matrix where 'i' and 'j' representing row and column of the matrix. Therefore, using Eqs. (13.2) and (13.3) we can derive Eq. 13.4.

$$\frac{\partial c}{\partial d^{\mathrm{T}}} = \frac{\partial c}{\partial b^{\mathrm{T}}} \cdot \frac{\partial b}{\partial d^{\mathrm{T}}} \tag{13.4}$$

CNN architecture is a combination of various layers and input that is tensors are processed from each layer in such a way that the output of one layer becomes the input for another layer. There are three major layers of the CNN model that are convolution, pooling, and fully connected layer as shown in Fig. 13.1.

Convolution Layer

Convolution layer is used for identifying the basic edges like horizontal and vertical. In deep convolution, various other more complex edges are identified. To understand

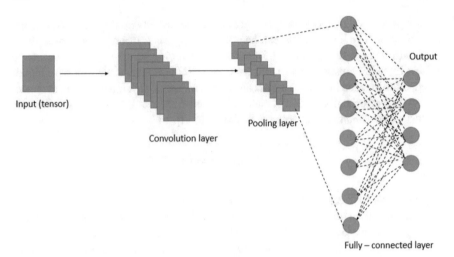

Fig. 13.1 Convolution neural network

the working of the convolution layer, let us consider a 3 × 4 input image and a 2 × 2 convolution kernel. We overlap the convolution kernel over the image and calculate the product as shown below in Fig. 13.2.

The convolution kernel is placed over the right-most corner of the input image and the product is calculated. Then it shifted one pixel down to calculate the new product and the process recursively followed till the kernel matrix reach the bottom pixels of the image as shown in Fig. 13.3. In Fig. 13.3, the summing of the products is done from the right top to the bottom. When the kernel matrix is placed over the

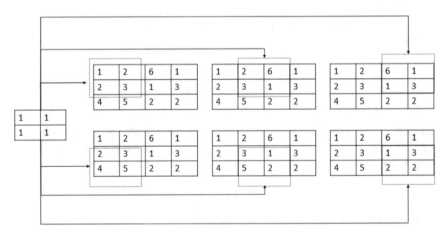

Fig. 13.2 2 × 2 convolution kernel is overlapped over 3 × 4 input image

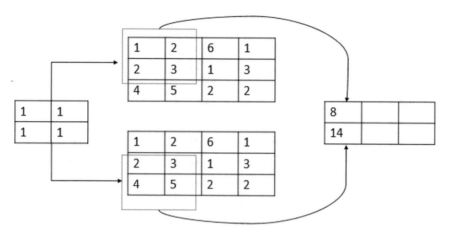

Fig. 13.3 Summing the products from top right to bottom

right top then the resultant of the sum of the product of overlapping pixels will be $(1 \times 1) + (1 \times 2) + (1 \times 2) + (1 \times 3) = 8$. Similarly, the kernel matrix is moved one pixel down and the sum of product is calculated for the overlapping pixel that is $(1 \times 2) + (1 \times 4) + (1 \times 3) + (1 \times 5) = 14$.

Now, we move the convolution kernel matrix to the top and calculate the SOP of overlapping pixels which would be $(1 \times 2) + (1 \times 3) + (1 \times 6) + (1 \times 1) = 12$. And, now the SOP is calculated for below window that is $(1 \times 3) + (1 \times 5) + (1 \times 1) + (1 \times 2) = 11$. This process continues for the remaining pixels also as shown in Figs. 13.4 and 13.5, respectively. In this way, after calculating the SOP from the right top corner of the input image and shifting the kernel window to the bottom, and again starting it from the next top to bottom we get the final resultant matrix as shown in Fig. 13.5.

Edge detection is one of the major issues in brain tumor detection. Convolution network provide various filters of kernels for detecting almost every edge. Majorly in convolution network, these kernels are of size of $3 \times 3 \times 3$. More specifically filters for detecting or creating feature maps highlighting the details of horizontal lines (KH) and vertical lines (KV) in the image shown (Fig. 13.6).

Pooling Layer

The output of the convolution layer is given to the pooling, pooling is applied to each feature map separately for reducing the size of feature maps, fast computation, and reduce the requirement of the memory. By reducing the number of parameters, pooling control the overfitting [17]. The most general form of pooling is the use of a 2D filter pass over the output of the convolution layer that is over each feature map to reduce its size. Pooling can be broadly categorized into three types max, average

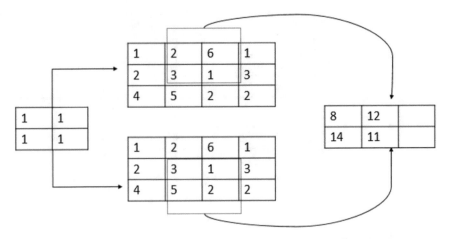

Fig. 13.4 Calculating SOP of overlapping pixels from middle top to bottom of input image

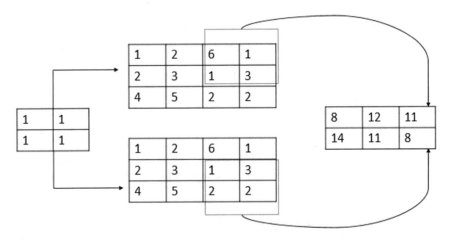

Fig. 13.5 Calculating SOP for last top to bottom and the final matrix

Fig. 13.6 Kernels for detection horizontal and vertical edges

$$K_H = \begin{bmatrix} 1 & 1 & 1 \\ 0 & 0 & 0 \\ -1 & -1 & -1 \end{bmatrix} \qquad K_V = \begin{bmatrix} 1 & 0 & -1 \\ 1 & 0 & -1 \\ 1 & 0 & -1 \end{bmatrix}$$

and global pooling. In the case of max pooling, the 2D filter passes over each feature map to reduce it by taking the maximum value among the values of the feature map within a block as shown in Fig. 13.7.

Where a block slides over a 4 × 4 feature map to downsample the feature map to 2 × 2. In the case of average pooling, instead of taking a maximum value among the values lie in a block sliding over a feature map an average of all values is taken. As

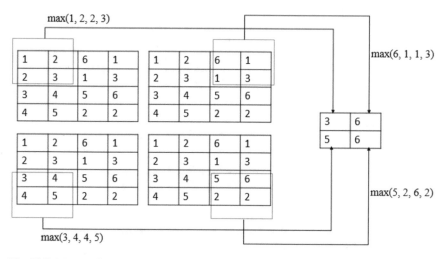

Fig. 13.7   Max pooling

shown in Fig. 13.8, the average of all values of the feature map that are coming inside the block is taken. Pooling is applied on all the feature maps separately so the size of each feature map gets reduced. Max pooling is used when the "less important" feature is not at all matters so in max pooling we only consider the maximum value of the feature map under the sliding block. Whereas average pooling can be the better option when we don't want to completely neglect the "less important" features. The major difference between the max and average pooling is that in the case of max-pooling only the most strong feature is considered and the rest of all features are neglected, but this is not happening in the case of average pooling.

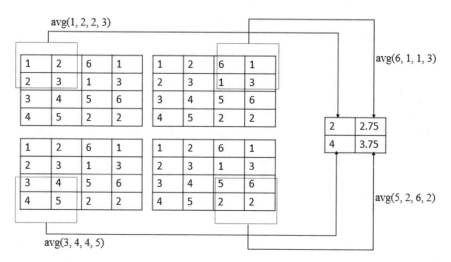

Fig. 13.8   Average pooling

Fig. 13.9 Global max
pooling

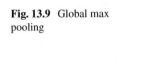

Fig. 13.10 Global average
pooling

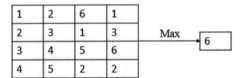

Global pooling can be categorized as global max and global average pooling, global pooling further downsample the feature map. Global pooling reduces each feature map to a single value, in the case of GMP feature map is replaced by the maximum value among the whole feature map. And, in GAP feature map is replaced by the average of all the values of the feature map as shown in Figs. 13.9 and 13.10.

Fully Connected Layer

The output of the pooling layer is flatten into a column vector. Suppose that resultant volume of max pooling is $3 \times 3 \times 100$ it means 100 matrices of 3×3 dimension. Then the output volume is flatten into column vector of dimension 1×900 and given as input to the fully connected layer of convolutional network. The number of neurons in the input layer would be equal to the length of column vector (900 in this case). These inputs passed with multiple hidden layers and activation functions before reaching to the output layer after first forward propagation step as shown in Fig. 13.11. The green boxes are hidden layers with number of neurons. Input layer fully connected with first hidden layer. That means every neuron of input layer is connected with each neuron of first hidden layer like, first neuron as identified in the figure is connected with every neuron of first hidden layer. All hidden layers are also fully connected. Similarly, the last hidden layer remain fully connected with the output layer. To keep figure simple, full connection between the various layers shown with the single arrowed line.

Application of CNN in Brain Tumor Segmentation

Brain tumor can be broadly classify in two categories Malignant and Benign [18]. Malignant tumor show fast division of cells and grow exponentially in very less time.

Fig. 13.11 Fully connected
layer of CNN model

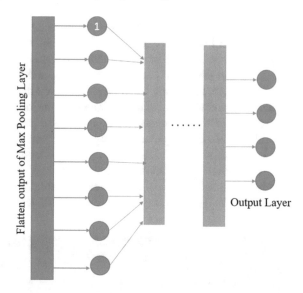

Malignant tumors are life threatening therefore brain tumor segmentation mainly focus on segmenting the malignant tumors. Due to the abnormal growth and irregular shape of brain tumor, it is always challenging to segment it with high accuracy. MRI play pivotal role in analysis of brain tumor and providing images of brain with different modalities. Basic modalities of brain tumor are FLAIR, T1, and T2 as shown in Fig. 13.11 [19]. Based on these MRI images, the machine operator and radiologist prepare a report about the present state of tumor. This manual segmentation of brain tumor is very tedious task and depends on the knowledge and experience of the operator. This manual segmentation always remain prone to human errors and non-uniformity between reports of two different machine under different environment. The report of radiologist considered as base for any doctor to start the medication of a patient. Moreover, the correct treatment of brain tumor highly depends on the report of MRIs.

To overcome all the above drawbacks, it is very much required to have an automated method for segmenting the brain tumor with high accuracy based on MRIs. Therefore, in past decades many researchers has proposed multiple methods and state-of-arts to give an efficient way to automate this process of manual segmentation and provide segmentation of all lesions with high accuracy.

Till 2015 machine learning algorithm like SVM [20], logistic regression [21], random forest [22], KNN [23], and K-mean clustering [24] are highly used by the researchers to proposed various semi-automated and automated methods of brain tumor segmentation. There machine learning methods are highly efficient and require very less computation time. These methods were preferred as they can give better results on limited amount of data [25–27]. However, it observed that even with increasing the training data the accuracy of these methods remain almost constant as shown in Fig. 13.12. Moreover, machine learning methods require handcraft feature extraction to train the model [28–31]. The result of any machine learning based model remain highly dependent on the capability of extracting features from the training images. Some efficient methods like GLCM [32], run-length encoding [33], LBP [34], and HMM [35] are used by the researchers for extracting features from training images. Extracting handcraft features are the images is the tedious task and increase the probability of skipping important information. Deep learning algorithm almost completely overcome the two main disadvantage of using machine learning algorithm. Deep learning methods improve the accuracy with increasing the training data as shown in Fig. 13.13. Moreover, do not require any pre-processing process of feature extraction before training the model. With these big advantages of deep learning approaches, in last decade research area of brain tumor segmentation get inundated with deep learning-based research articles.

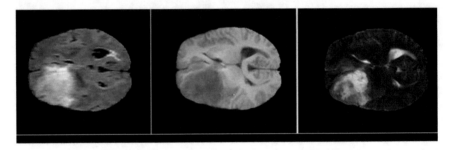

Fig. 13.12 Different modalities of brain tumor in MRI

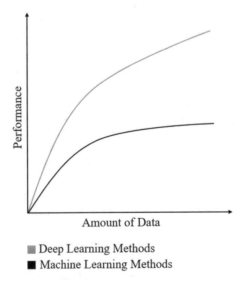

Fig. 13.13 Change in performance of machine learning and deep learning based methods with increasing amount of data

Multiple researchers used conventional CNN method and also propose various novel architectures, based on conventional CNN for automatic segmentation of brain tumor with higher accuracy. Almost every researcher use BraTS dataset for training and testing purpose. Summary of few studies mentioned in the upcoming lines. Wang et al. [36] used convolutional neural network with multiple convolution and pooling layers for segmenting the brain tumor. The experiments carried out using BraTS 2015 [37, 38] dataset and shows classification accuracy of 83%. Pereira et al. [39], proposed a novel hierarchical brain tumor segmentation model to automate the process of segmenting lesions in MRI images. The proposed model was based fully convolutional neural network as shown in Fig. 13.14, in which overall tumor is segmented first. Moreover, intra-tumors segmented for better precision and accuracy. Experimental study done using BraTS 2013 dataset for training, validation, and testing purpose of the model. The results shows better Dice Score Coefficient (DCS) $0.85 \pm 0.05, 0.76 \pm 0.17, 0.74 \pm 0.08$ for segmenting complete, core, and enhancing tumor.

Zhao et al. [22] combine fully convolutional model with conditional random fields to propose a novel architecture for automatic segmentation of brain tumor. Conducted experiments on BraTS 2013 dataset and shown improved dice score of 0.87, 0.83, and 0.78 for complete, core, and enhancing tumor segmentation. Hussain et al. [40] used deep convolutional neural network model to conduct patch-based approach to train a model. The training images are pre-processed and divided into various patches to provide input to CNN and the output of CNN post-processed to get better results. Conducted experimental study using BraTS 2013 and 2015 dataset shows improved accuracy as compare to other state-of-art methods.

Conclusion

In this chapter, we discuss the use of MRIs for the doctors to diagnose the brain tumor, and how important is to provide an automatic method for segmenting brain tumor in the reports of MRIs. Deep learning algorithm has great importance in providing various ways for automatically detecting the tumor cell in the brain MRIs. CNN is the most common and widely used algorithm of deep learning used by various researchers to solve the purpose. Therefore, we discuss convolutional neural network method mainly highlighting the process of convolution and pooling. Also, mentioned the application of convolutional neural network based proposed architectures for segmenting the brain tumor using MR images.

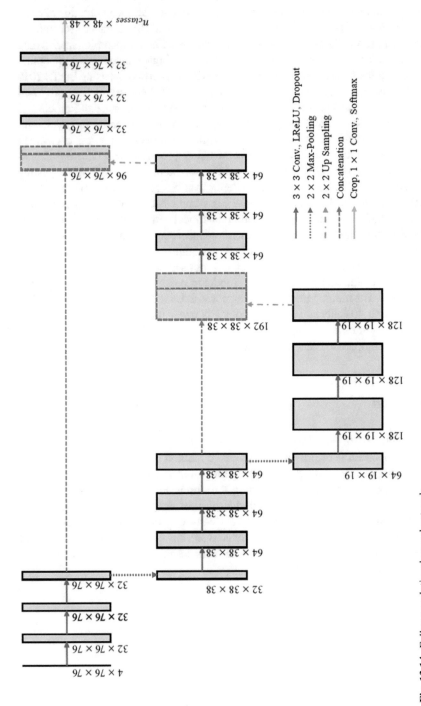

Fig. 13.14 Fully convolutional neural network

References

1. American Society of Clinical Oncology (ASCO). http://www.asco.org/
2. John, P.: Brain tumor classification using wavelet and texture based neural network. Int. J. Sci. Eng. Res. **3**(10), 1–7 (2012)
3. Alfonse, M., Salem, A.B.M.: An automatic classification of brain tumors through MRI using support vector machine. Egy. Comp. Sci. J. **40**(3) (2016)
4. Pereira, S., Pinto, A., Alves, V., Silva, C.A.: Brain tumor segmentation using convolutional neural networks in MRI images. IEEE Trans. Med. Imaging **35**(5), 1240–1251 (2016)
5. Dandıl, E., Çakıroğlu, M., Ekşi, Z.: Computer-aided diagnosis of malign and benign brain tumors on MR images. In: International Conference on ICT Innovations, pp. 157–166. Springer, Cham (2014)
6. Xiao, Z., Huang, R., Ding, Y., Lan, T., Dong, R., Qin, Z., et al.: A deep learning-based segmentation method for brain tumor in MR images. In: 2016 IEEE 6th International Conference on Computational Advances in Bio and Medical Sciences (ICCABS), pp. 1–6. IEEE (2016)
7. Balasubramanian, C., Saravanan, S., Srinivasagan, K.G., Duraiswamy, K.: Automatic segmentation of brain tumor from MR image using region growing technique. Life Sci. J. **10**(2) (2013)
8. Selvakumar, J., Lakshmi, A., Arivoli, T.: Brain tumor segmentation and its area calculation in brain MR images using K-mean clustering and Fuzzy C-mean algorithm. In: IEEE-International Conference on Advances in Engineering, Science and Management (ICAESM-2012), pp. 186–190. IEEE (2012)
9. Clarke, L.P., Velthuizen, R.P., Camacho, M.A., Heine, J.J., Vaidyanathan, M., Hall, L.O., et al.: MRI segmentation: methods and applications. Magn. Reson. Imaging **13**(3), 343–368 (1995)
10. Pouyanfar, S., Sadiq, S., Yan, Y., Tian, H., Tao, Y., Reyes, M.P., et al.: A survey on deep learning: algorithms, techniques, and applications. ACM Comput. Surveys (CSUR) **51**(5), 1–36 (2018)
11. Pereira, S., Alves, V., Silva, C.A.: Adaptive feature recombination and recalibration for semantic segmentation: application to brain tumor segmentation in MRI. In: International Conference on Medical Image Computing and Computer-Assisted Intervention, pp. 706–714. Springer, Cham (2018)
12. Najafabadi, M.M., Villanustre, F., Khoshgoftaar, T.M., Seliya, N., Wald, R., Muharemagic, E.: Deep learning applications and challenges in big data analytics. J. Big Data **2**(1), 1–21 (2015)
13. Gao, X.W., Hui, R., Tian, Z.: Classification of CT brain images based on deep learning networks. Comput. Methods Programs Biomed. **138**, 49–56 (2017)
14. Zhao, L., Jia, K.: Deep feature learning with discrimination mechanism for brain tumor segmentation and diagnosis. In: 2015 International Conference on Intelligent Information Hiding and Multimedia Signal Processing (IIH-MSP), pp. 306–309. IEEE (2015)
15. Wu, J.: Introduction to convolutional neural networks. National Key Lab for Novel Software Technology, vol. 5, p. 23. Nanjing University, China (2017)
16. Stock, K., Pouchet, L.N., Sadayappan, P.: Using machine learning to improve automatic vectorization. ACM Trans. Architecture Code Optimization (TACO) **8**(4), 1–23 (2012)
17. O'Shea, K., Nash, R.: An introduction to convolutional neural networks. arXiv preprint arXiv: 1511.08458 (2015)
18. Rezaeil, K., Agahi, H.: Malignant and benign brain tumor segmentation and classification using SVM with weighted kernel width. Signal Image Process. Int. J. (SIPIJ) **8**(2), 25–36 (2017)
19. Bakas, S., et al.: Identifying the best machine learning algorithms for brain tumor segmentation, progression assessment, and overall survival prediction in the BRATS challenge. arXiv preprint arXiv:1811.02629 (2018)
20. Amin, J., et al.: Brain tumor detection using statistical and machine learning method. Comput. Methods Programs Biomed. **177**, 69–79 (2019)
21. Havaei, M., et al.: Within-brain classification for brain tumor segmentation. Int. J. Comput. Assist. Radiol. Surg. **11**(5), 777–788 (2016)
22. Zhao, X., et al.: A deep learning model integrating FCNNs and CRFs for brain tumor segmentation. Med. Image Anal. **43**, 98–111 (2018)

23. Işın, A., Direkoğlu, C., Şah, M.: Review of MRI-based brain tumor image segmentation using deep learning methods. Procedia Comput. Sci. **102**, 317–324 (2016)
24. Havaei, M., et al.: Brain tumor segmentation with deep neural networks. Med. Imag. Anal. **35**, 18–31 (2017)
25. Magadza, T., Viriri, S.: Deep learning for brain tumor segmentation: a survey of state-of-the-art. J. Imag. **7**(2), 19 (2021)
26. Suthaharan, S.: Support vector machine. Machine learning models and algorithms for big data classification, pp. 207–235. Springer, Boston, MA (2016)
27. Peng, C.-Y.J., Lee, K.L., Ingersoll, G.M.: An introduction to logistic regression analysis and reporting. J. Educ. Res. **96**(1), 3–14 (2002)
28. Livingston, F.: Implementation of Breiman's random forest machine learning algorithm. ECE591Q Mach. Learn. J. Paper 1–13 (2005)
29. Bijalwan, V., et al.: KNN based machine learning approach for text and document mining. Int. J. Database Theory Appl. **7**(1), 61–70 (2014)
30. Ahmad, A., Dey, L.: A k-mean clustering algorithm for mixed numeric and categorical data. Data Knowl. Eng. **63**(2), 503–527 (2007)
31. Mohanaiah, P., Sathyanarayana, P., GuruKumar, L.: Image texture feature extraction using GLCM approach. Int. J. Sci. Res. Publ. **3**(5), 1–5 (2013)
32. Bradley, S.D.: Optimizing a scheme for run length encoding. Proc. IEEE **57**(1), 108–109 (1969)
33. Guo, Z., Zhang, L., Zhang, D.: A completed modeling of local binary pattern operator for texture classification. IEEE Trans. Image Process. **19**(6), 1657–1663 (2010)
34. Miller, David, R.H., Leek, T., Schwartz, R.M.: A hidden Markov model information retrieval system. In: Proceedings of the 22nd Annual International ACM SIGIR Conference on Research and Development in Information Retrieval (1999)
35. Angulakshmi, M., Lakshmi Priya, G.G.: Automated brain tumour segmentation techniques—a review. Int. J. Imaging Syst. Technol. **27**(1), 66–77 (2017)
36. Wang, F., et al.: The application of series multi-pooling convolutional neural networks for medical image segmentation. Int. J. Distrib. Sensor Netw. **13**(12), 1550147717748899 (2017)
37. Menze, B.H., Jakab, A., Bauer, S., et al.: The multimodal brain tumor image segmentation benchmark (BRATS). IEEE T Med. Imaging **34**(10), 1993–2024 (2015)
38. Kistler, M., Bonaretti, S., Pfahrer, M., et al.: The virtual skeleton database: an open access repository for biomedical research and collaboration. J. Med. Internet Res. **15**(11), e245 (2013)
39. Pereira, S., et al.: On hierarchical brain tumor segmentation in MRI using fully convolutional neural networks: a preliminary study. In: 2017 IEEE 5th Portuguese Meeting on Bioengineering (ENBENG). IEEE (2017)
40. Hussain, S., Anwar, S.M., Majid, M.: Segmentation of glioma tumors in brain using deep convolutional neural network. Neurocomputing **282**, 248–261 (2018)

Chapter 14
Transportation Management Using IoT

Deep Learning to Predict Various Traffic States

Amit Singh

Introduction

Transporting general-purpose commodities in public and private vehicles is one of the biggest tasks that need to be planned properly with the planning and monitoring authorities. The road traffic in due increasing public and private vehicles are also congesting roads and thus, the meeting with accidents and stuck in blockage is common. Depending on the grade of the services acquire, the planning and monitoring authorities has to decide an optimal solution that satisfies the multi-constraints transportation problem. Figure 14.1 shows the multiple dimensions involved and related to the transportation management system from the manufacturing of the product to delivery to the consumer.

Such issues related to the transportation system are faced around the globe. The dynamics of societal changes in the context of the rail transportation system of the Netherland are discussed in [1]. The challenges and market demand of world-class rail transportation in terms of passengers and goods are highlighted in the article. Tyan et al. discussed collaborative transport management (CTM) for e-commerce facilities to reduce the delivery time and reliability [2]. A heuristic optimization-based algorithm with critical parameters, such as safety and environmental cost is considered in [3] for logistic management supports in Spain. To reduce road congestion, pollution, and accidents from the Indian perspective, an article encouraging public transportation and suppressing private vehicles is proposed in [4].

A. Singh (✉)
School of Computer Science, University of Petroleum and Energy Studies, Dehradun, India
e-mail: amit10jul1980@gmail.com

© The Author(s), under exclusive license to Springer Nature Singapore Pte Ltd. 2023
V. Kadyan et al. (eds.), *Deep Learning Technologies for the Sustainable Development Goals*, Advanced Technologies and Societal Change,
https://doi.org/10.1007/978-981-19-5723-9_14

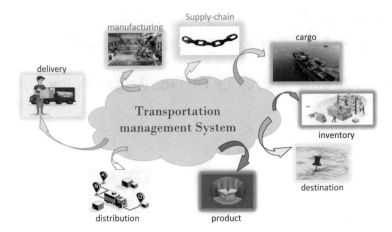

Fig. 14.1 Components of transport management system

The transportation challenges are not untouched in developed countries as well. Msigwa presented the challenges and remedial solutions for Tanzania and highlighted the issues, such as accidents, air pollution, traffic congestion, and energy consumption with suggestive measures to reduce it [5].

The success of any business depends on the transportation management system that how efficiently and timely it is transferring its goods from warehouse to consumers' places. Doing manually, tasks such as order segmentation, route planning, and vehicle mapping with orders cannot guarantee precision in work. This manual mechanism of the twenty-first century may lead to pitfalls in terms of business performance and profit. Existing literature shows a huge number of articles demonstrating and addressing the challenges in transportation management in the context of public and goods transportation through air, road, and water [6]. Based on the real-time issues during the shipment of goods, a few of the challenges of transportation management are listed in Fig. 14.2.

Fig. 14.2 Issues with the traditional transportation system

Transportation challenges:
- Optimality
  - □ Unused vehicle capacity
  - □ Shortest distance shipping path
  - □ Transportation cost
- Customer curiosity
  - □ Consignment tracking
  - □ Inconvenient order delivery
- Inventory management
  - □ Logistic support
  - □ Mismanaged inventory

Depending upon the type of consignments and way of transportation and delivery, the issues vary from one to many. The above-mentioned issues are as follows.

Under-Utilization of Vehicle Capacity

The conventional way of managing transportation systems sometimes does not estimate the proper information of vehicle capacity and order-vehicle mapping, etc. Moreover, when the system is too large, the outcome always falls due to the under-utilization of resources. Such shortfalls degrade the efficiency in all sectors, such as goods transportation, passenger vehicles of road, water, and air transportations.

Route Optimality

The real-time traffic needs to be shared with the vehicle drivers so that the runtime decisions can be taken to avoid road congestions and rutted rough roads. The shortest route may not be always a good option due to other factors that can accumulate delay in the destination.

Order Tracking

Especially tracking of consignment at run time is another challenging issue because once a consignment is left from the warehouse it is no more in our surveillance. Hence, using a customary transportation management system requires a huge number of human resources to connect with the vehicle staff over multiple calls to get an update.

Untimely Delivery of the Consignment

Another woe to the transportation system is a due cause of delay in scheduled delivery of goods. Some goods, such as vegetables and fruits have a very short expiry period and need to be dispatched soon up to the consumer's place. This kind of challenge directly impacts customer satisfaction and experience.

Low Visibility of Inventory and Logistics

Human resources can manage the thing, however, the automatic system can be better perform the job. The logistics and inventory control can be enhanced to reduce the manpower for the same.

Transportation Cost

The cost depends on the efficient shipping of goods. How many human resources are required, what needs to be transported how fast to the destination, and how the inventory needs to be updated to ensure supply-chain management.

The conventional transportation management system has immeasurably more challenging issues that can be addressed through the use of technology. The recent advancement in sensor and RFID technologies has garnered attraction in developing smart devices and objects of the real world. With IoT, physical devices can interact with each other smartly and communicate their data for further applicability [7].

IoT in Transportation Management

The concept of structuring smart cities and smart buildings is based on an IoT system. Any physical device, such as houses, vehicles running on the road, roadside poles, trees, etc. can be IoT-equipped [8]. Irrespective of static and dynamic, IoT-equipped objects can share their runtime information with other objects and thus, can communicate to each other. For example, if cars running on the road can communicate with each other, it may utilize the information by other cars' drivers to avoid congestion and rutted rough roadway. Road traffic is analyzed in the context of transportation management in [9]. The author termed it as "social transportation" to this vehicle-to-vehicle communication during run time in the traffic.

Communication among IoT-enabled devices is carried out via the internet [10, 11]. Although the usage of the internet in IoT devices has its challenges that need to be addressed and is another dimension of the study. However, transportation uses commercial public and private vehicles to deliver products from manufacturing to the warehouse, warehouse to the dealer, dealer to the retailer, and finally to the consumer. Simultaneously, transportation systems are used for routine human commute purposes. If these passengers and goods vehicles are equipped with IoT technology that can share the runtime data with another vehicle via the Internet, transportation-related issues mentioned above can be addressed to a greater extent.

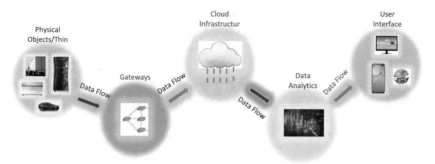

Fig. 14.3 Elements of Internet of Things

IoT Components and Information Accomplishment

IoT module is a combination of separate electronic components, such as sensors/ actuators, radio frequency identification system (RFID), software for information processing, etc. The IoT processing components using which any smart object can communicate to another nearby smart object are shown in Fig. 14.3.

As depicted in Fig. 14.2, the physical objects are equipped with different types of sensors that can collect the data from the surrounding environment and transfer it to the cloud infrastructure through gateways. The data received at the cloud can be analyzed for different applications as per the requirement of the user. The result obtained can further be transported to the user interface for decision-making. The steps involved in the IoT-enabled physical objects for collecting environmental analog data, processing it for information, and suggesting users for appropriate resolution.

The IoT technology collects the data and sends it to the cloud for processing in the desired output forms. The processing of exceedingly large data is not an easy task in the cloud, and hence, deep learning-based neural network technologies are an efficient solution to process unlabeled or unstructured data. Deep learning [12, 13] is a branch of machine learning that covers and discusses the algorithms motivated by the organization and activity of the brain popularly known as artificial neural network (ANN) [14, 15]. Deep learning applies to raw data over multiple layers to extract the higher-level of features. This chapter explores the depth and breadth of deep learning technologies and their applications over the data collected by IoT-equipped transportation management.

IoT Dimensions in Transportation Management

As depicted in Fig. 14.1, manufacturing, inventory, delivery, product, shipping, etc. are the few important dimension of a business process. The transportation of goods

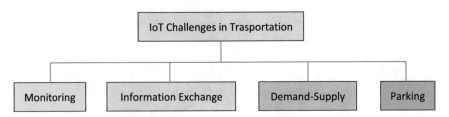

Fig. 14.4 IoT in transportation: Challenges

from the production unit to vendor warehouses of different cities, states, and countries can be monitored through IoT to enhance transparency and surveillance throughout the way. Managing passenger vehicles over the roads is another summons on it. Though the IoT application in the transport management system is very wide, a few of them are highlighted as in Fig. 14.4.

Monitoring and Information propagation in runtime. Tracking of in-transit goods and any dear ones in real time is one of the IoT measures in the context of transportation management. Hence, the information propagation to the respective station becomes easy and improves business strategies, and thus helps in turns to the traffic department. Monitoring and control of fluid transportation are presented in [16] and data is analyzed by machine learning. An IoT-equipped smart car is monitored using predictive maintenance in [17] for optimal fuel usage exhaust emissions.

Managing Supply-chain [18, 19]. Being a complex process, supply-chain requires fast communication, adaptability against changes, risk management, and understanding among stakeholders in runtime. In addition, the availability of resources can easily be tracked and effective fulfillment of the actual requirements can be achieved using IoT deployment. The oversupply and shortage of demand can also be reduced through an IoT-equipped transportation management system. A supply-chain the traffic information ahead on the road and thus, the diversions can be explored by the vehicles risk and case reasoning is exposed in [20] using IoT information, such as 3G, RFID, and GPS. The simulation results show better performance. When it comes to insufficient or lack of information, the Decision-Making Trial and Evaluation Laboratory (N-DEMATEL) technique with analytic hierarchy process (AHP) in the neutrosophic environment is presented and validated against the traditional supply-chain management [21].

Traffic avoidance and re-routing [22]. Vehicle-to-vehicle communication shares coming behind to avoid being stuck in a jam. Masek et al. [23] presented an IoT-driven environment for traffic modeling. Similar work for parallel transportation management is proposed in [24] to control and manage road traffic. A detailed review of static and dynamic traffic light systems (TLS), RFID, and IoT is explored by Avatefipour and Sadry [25] for intelligent transportation management (ITS).

Parking control and care [26–28]. The number of vehicles on the road is increasing every day and thus, the challenges of parking space and security are rising as the biggest problem for smart cities. Abdulkader et al. [29] presented a secure and controlled parking solution using the integration of wireless sensor networks, RFID, and IoT technologies. A real-time parking lot reservation with an e-payment solution is targeted and obtained to avoid congestion and enhance user satisfaction. Experimentation based on parking location, days of the week, and hours of the day is performed in [30] using deep long-short memory networks. Birmingham parking sensors dataset is used for the prediction of parking space using IoT-cloud infrastructure and sensory network.

To address such issues, the machine learning-based prediction models are very helpful to handle large datasets. Human nerves system-based artificial neural network and with inspired deep learning models figure out some extent.

Basic ANN and Deep Learning [31–33]

Deep learning technologies are evolved from artificial neural network (ANN) and vary as deep neural networks (DNN), deep belief networks (DBN), deep reinforcement learning (DRL), recurrent neural networks (RNN), and convolutional neural networks (CNN). The application domain of deep learning is not limited to computer vision, natural language processing, speech recognition, machine translation, and healthcare and analytics. The unbounded number of multiple layers of bounded length in deep learning is more application-oriented and optimized against the simplest ANN of a single layer of unbounded size. The pictorial representation of multilayer (2-hidden layer) and simplex (1-hidden layer) ANN is shown in Fig. 14.5.

Figure 14.5a is an artificial neuron; an imitation of a human neuron structure that is used to build an artificial neural network. Interconnection (i.e., weight) of such artificial neurons having proper threshold (i.e., bias) collectively form an ANN structure shown in Fig. 14.5b, c. The input layer receives the labelled or unlabeled data, while the processed data and desired results are obtained through output layer. With the variation in the number of hidden layers, the efficiency of ANN for nonlinear data sets increases. An ANN with more than 1-hidden layer considers a deep neural network.

Deep Learning Using IoT in Transportation

Deep learning has been observed great success in the field of natural language processing, recommendation system, computer vision, and the audio recognition domains. In the context of the transportation management system, usage of IoT for data collection and further converting into data sets for deep neural networks can measuredly enhance transportation performance. An exhaustive survey of deep

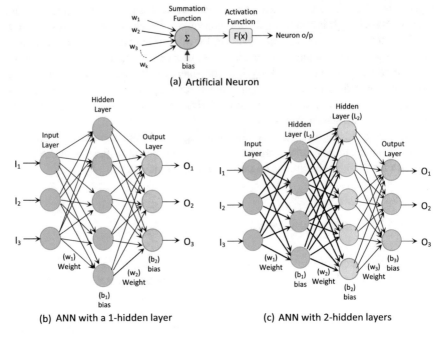

Fig. 14.5 Artificial neural network structure

learning techniques in transportation is presented in [34] that explores the basic DNNs, such as multi-layer perceptron (MLP), deep belief network (DBN), and sparse auto-encoder (SAE).

Some of the prediction models are presented in the following sections for traffic spatial and temporal traffic dependencies. The validation of prediction performance is measured through mean relative error (MRE), root means square error (RMSE), mean absolute error (MAE), and mean absolute percentage error (MAPE) as shown in Eqs. (14.1)–(14.4):

$$MRE = \frac{1}{N} \sum_{i=1}^{N} \frac{\left| Y_i - \hat{Y}_i \right|}{Y_i}. \tag{14.1}$$

$$RMSE = \sqrt[2]{\left[\frac{1}{N} \sum_{i=1}^{N} \left(Y_i - \hat{Y}_i \right)^2 \right]} \tag{14.2}$$

$$MAE = \frac{1}{N} \sum_{1}^{N} \left| Y_i - \hat{Y}_i \right|. \tag{14.3}$$

$$\text{MAPE} = \frac{1}{N} \sum_{1}^{N} \frac{\left| Y_i - \hat{Y}_i \right|}{Y_i} \qquad (14.4)$$

where

N number of sample tests
Y_i actual traffic data received from IoT devices
$\hat{Y}_i$ model prediction traffic flow.

These parameters are worn to measure and compare the accuracy of the prediction models. The subsequent section enlights a few traffic states, such as flow, vehicle speed, and travel time that are predicted through real-time traffic or simulation models.

Traffic Flow Prediction

Traffic flow prediction is a problem of the time series category that predict future traffic trend based on the data collected over a while in the observed location.

Deep learning-based Restricted Boltzmann Machine (RBM). RBM is used to analyze the short-term traffic flow prediction [35]. The real-time multimedia traffic data aggregation is considered through the Internet of Vehicles (IoVs) to train the model. The experimental results show the performance enhancement over the traditional neural network. An actual short-term traffic flow of Los Angeles Expressway of 10 days duration srting from 3rd Apr. 2017 is used as sample data. The time interval of data collection is considered from 7:00 AM to 10:00 AM every 5 min. The accuracy of prediction is measured through MRE and RMSE as mentioned in Eqs. (14.1) and (14.2).

The result demonstrated by the authors is validated against the traditional frameworks and claimed the traffic flow forecasting close to the real-time traffic flow that is useful to control and manage the road traffic.

As shown in Table 14.1, the authors compared the performance against the traditional approaches, such as ANN, the back propagation (BP), the support vector regression machine (SVR), and the auto-regressive integrated moving average (ARIMA). The Restricted Boltzmann Machine (RBM) machine based on deep learning prediction for chaotic time series performs better than other traditional neural networks. The authors claimed the better prediction ability and nonlinear fitting for chaotic time series by increasing the training samples for RBM.

Light-weight DeepTrend 2.0. Another light-weighted model called DeepTrend 2.0 is proposed in [36] which is fundamentally inherited from the detrending and deep learning concept. Authors claim that the convolutional neural network (CNN)-based DeepTrend 2.0 outperformance the Graph-CNN (GCNN). The data has been

Table 14.1 Prediction accuracy of deep learning based RBM [35]

Metrics	Training samples	ANN	BP	SVR	ARMA	Deep learning based RBM
MRE	210	0.97	0.98	0.96	0.89	**0.84**
	220	1.04	0.95	0.93	0.84	**0.79**
	230	0.94	0.91	0.89	0.78	**0.71**
	240	0.92	0.90	0.86	0.76	**0.67**
	250	0.98	0.87	0.84	0.69	**0.63**
	260	0.95	0.83	0.81	0.67	**0.59**
RMSE	210	8.42	7.56	7.73	7.11	**6.32**
	220	8.05	7.94	7.56	6.89	**6.14**
	230	7.86	7.2	7.5	6.32	**5.88**
	240	7.74	7.32	6.83	5.84	**5.71**
	250	7.54	6.98	6.64	5.92	**5.43**
	260	7.1	6.84	6.71	6.07	**5.27**

collected from multiple sensors during experimentation for many-to-many predictions. The authors made use of residual traffic data instead of original traffic flow which is called detrending and includes the following three steps [37]:

(i)   Computation of trends based on the collected data
(ii) Classification of trends captured in previous step based on the residual time series
(iii) Developing a prediction model for the residual time series.

The authors use the simple average method to compute the traffic trends, i.e., the average of the periodic traffic flow at each sensor.

The model is experimented and validated through the evaluation metrics, such as RMSE, MAE, and MAPE for traffic flow prediction.

The authors claim the importance of pre-processing of traffic data through detrending to magnify the trends and subsequently enhance the traffic forecasting performance using deep learning technologies. Further, the authors claim a balanced trade-off between model complexity and accuracy during complex deep learning-based traffic flow predictions.

A short-term traffic prediction is presented by Polson et al. [38] for the impulsive traffic flow during Chicago football games and snowstorm events. The model is trained through basic input–output pair $(y_i, x_i)_{i=1}^{N}$ to minimize the difference between the training target (y_i) and forecasted value $(y(x_i))$. As a preprocessor, a median filter is used to reduce the physical noises observed during sensor data collection. The authors computed the residual, i.e., the difference of actual and forecasted traffic flow $\left(r_i = Y_i - \hat{Y}_i\right)$ to measure the accuracy of prediction. This residual is used to improve the efficacy of the model. The prediction efficiency of the best linear model

(VARM8L) is slightly higher than the neural network with one hidden layer. However, when it comes to the deep learning model (DLM8L) the prediction performance gain is 14%.

Hybrid DNN-Based Traffic Flow prediction (DNN-BTF). A hybrid approach based on CNN for spatial features and RNN for temporal feature collection of data traffic is presented in [39]. An open-dataset PeMS is used to validate the DNN-BTF on long-term traffic prediction. The dominant characteristics of CNN for spatial properties of images and videos are exploited in this association, whereas the influencing capability of RNN for temporal correlation is used.

The accuracy of the model is validated for least absolute shrinkage and selection operator (LASSO), BP neural network, SAE, DNN-based prediction model for spatial–temporal data (DeepST), and StoS with evaluation metrics MAE, MRE, and RMSE. The authors claim the best performance of DNN-BTF in terms of average correlation. The error analysis of different conditions, like time-of-day, day-of-week, and spatial location is considered to study the influence over the prediction model.

Sequence to Sequence Learning with Attention mechanisms (Seq2Seq-Att). The encoder-decoder-based RNN is first coined by Cho et al. [40], which was further refined in terms of learning [41] and subsequently considered as the prototype of the Se2Seq learning model. The Seq2Seq model is shown in Fig. 14.6 consists of a series of RNN/LSTM.

As depicted in Fig. 14.6, the sequence of the RNN/LSTM network is used to build sequence to sequence framework, where the left and right part of the model represents the encoder and decoder respectively. The encoder generates the fixed-sized vector from the inputs; feed to the model and the decoder system is used to visualization of the output. The fixed-sized vector is fed to the input of the initial layer of the decoder system to influence the outcome of the decoder system.

The application of the Seq2Seq learning model along with the attention mechanism is demonstrated for short-term passenger flow prediction in large-scale metro systems is demonstrated and analyzed in [42]. The incorporation of the attention

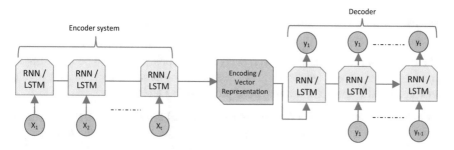

Fig. 14.6 Sequence-to-sequence learning model framework [40]

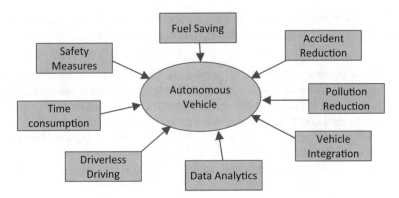

Fig. 14.7 Autonomous vehicle research measures

mechanism is to focus on the particular part of the input dataset to enhance the prediction performance. The experimental results are validated against the conventional deep learning approaches for significant performance improvement.

As the number of autonomous vehicles is increasing in recent years, research direction is now focusing in that direction for different kinds of challenges. Another interesting segment of research for traffic flow prediction is reviewed in [43] for autonomous vehicles. The following parameters shown in Fig. 14.7 are considered while developing deep learning techniques for traffic flow prediction. An efficient traffic flow prediction can obtain the runtime route guidance, reduction in traffic congestion, travel onboard decision-making, time-saving, etc.

Deep learning-based CNN, RNN, LSTM, GRU, Time Delay NN (TDN), DBM, RBM, DBN, and Wavelet Neural Network (WNN) in the field of the autonomous vehicle is extensively reviewed with their pros and cons.

Futuristic traffic flow prediction through the previously collected historical data is required to avoid overcrowding and blockage on roads. Hence, ease the traffic flow.

Traffic Speed Prediction

The forecasting of the number of vehicles passing a point at a given point of time is known as traffic speed prediction. Few deep learning-based approaches are highly recommended in the literature for speed prediction as given below:

Path-based deep Learning framework for Critical Path (PBDL-CP). A spatial–temporal traffic state is addressed in [44] using a path-based deep learning framework. The road network is divided into critical paths (i.e. most frequently used paths) and each path is modeled through the bidirectional long short-term memory neural network (Bi-LSTM NN), and multiple Bi-LSTM layers. These two neural networks

Fig. 14.8 Path-based deep learning framework for road traffic speed prediction [44]

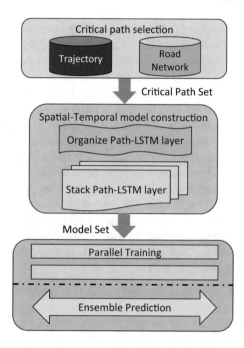

are used to gather the temporal information of the traffic which further, captures spatial–temporal information during predicting the traffic speed. The PBDL-CP framework is depicted in Fig. 14.8.

The critical path selection is performed through historical trajectories and road networks provided to the upper layer of the framework. These critical paths are analyzed by the middle layer with bidirectional LSMT neural networks for spatial–temporal information collection of different segments of the road network. Further, the information of spatial and temporal is used to train the model in parallel as shown in the lower layer.

The average MSE of PBDL-CP is reduced by 35.40%, 21.09%, 9.92%, and 16.90% for K-nearest neighbor (KNN), ANN, CNN, and LSTM NN, respectively, and subsequently, authors concluded the best prediction performer in the solution category. It has been demonstrated that with the increase of input timespans and prediction timespans, the deep learning-based model shows better-hidden layer utilization to solve non-linear complex problems. The road network is divided into segments and validated 112 segments dataset for MSE comparisons among different solution approaches. However, the prediction approximately obtained 86% segments during short-term and 70% segments of the long-term prediction and validated the applicability of the PBDL-CP model for urban traffic speed prediction.

Attention Graph Convolutional Sequence-to-Sequence model (AGC-Seq2Seq) [45]. A substantial effort has been encircled for multistep traffic speed prediction to address Spatio-temporal dependencies of traffic conditions through the Seq2Seq model and GCNN. Road topology is considered as a graph and the graph convolution captures the spatial information of the traffic. Further, the Seg2Seq model composed of two connected RNN (gated recurrent units for encoding and decoding operation) structures translates the spatial time series data into Spatio-temporal dependencies. Due to the discrepant distribution of input between training and testing timespan, a new training strategy is designed to coordinate the multidimensional features of Spatio-temporal traffic speed variables. Authors considered historical statistics traffic information and time-of-day as inputs, whereas historical information is used for training and testing stages and thusly, enhances the prediction accuracy.

The simulation results obtained during experimentation are validated on real-time dataset A-map in terms of measuring prediction error parameters, such as MAPE, MAE, and RMSE. Deep learning-based approaches obtained better prediction performance than the machine learning models but consume longer computation time. Though the model is concluded for the improved prediction performance against the Graph-CNN and Seq2Seq-Att, the time complexity of the proposed AGC-Seq2Seq is slightly higher.

Fusion deep learning [46]. It is modeled to address lane-level traffic speed. The entropy-based gray relational analysis is adopted for the selection of correlated lane sections to overcome the collection of high-quality data to describe the traffic state. Further, a two-layer deep learning framework is used for the prediction of traffic speed. The deep learning model uses the combination of LSTM and GRU neural networks for this purpose. Remote traffic microwave sensors (RTMS) are used for traffic data collection for the training, testing, and validation of the proposed work. Based on the MAE, RMSE, MAPE, and variance of absolute percentage error (VAPE) evaluation metrics, the experimental results are significantly improved against ARIMA, LWR, MLP, KF, RBFNN, CNN, and LSTM. VAPE is measures are per Eq. 14.5.

$$\text{VAPE} = \sqrt{\frac{L\sum_{i=1}^{N}\left(\frac{\left|\hat{Y}_i - Y_i\right|}{Y_i}\right)^2 - \left[\sum_{i=1}^{L}\left(\frac{\left|\hat{Y}_i - Y_i\right|}{Y_i}\right)\right]^2}{L(L-1)}} \qquad (14.5)$$

where L is the size of the prediction sequence. Unlike other measuring parameters, VAPE evaluates the stability of the prediction model rather than accuracy.

Multitask learning Model [47]. Another deep learning-based traffic prediction approach was introduced over the multitask learning model. The basis of this model is:

(i)   To enlighten the most informative features of the spatial–temporal relationship, an abstraction-based Granger causality analysis is used.
(ii) Computationally efficient Bayesian optimization is used to control the impact of influencing parameters.

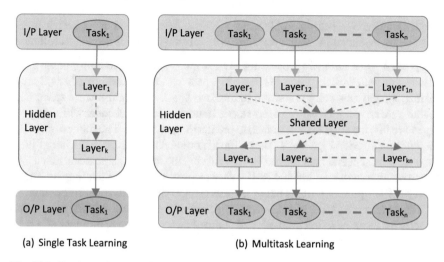

(a) Single Task Learning   (b) Multitask Learning

Fig. 14.9 Single- and multi-task architecture [47]

The authors claimed the prediction accuracy and efficiency for large-scale road networks. To capture Spatio-temporal patterns multitask layer is divided into two levels, unlike the single task learning model as shown in Fig. 14.9. In the first level, the model is trained for each road segment and amalgamate into a shared layer whereas, in the second level, the GRU is fed with the shared layer.

The proposed model MTL-GRU is compared with the STL-GRU, STL-LSTM, Conv-GRU, temporal convolutional network (TCN), and some classical methods, such as v-SVM, k-NN, and EFNN. The authors recorded a significant improvement in terms of MAPE against the mentioned approaches.

Geometric Deep Learning Approach [48]. City-wide traffic data is analyzed for speed prediction using a data-driven deep learning model. Graph convolution and attention mechanisms are used to capture geometric traffic data dependency. Encoder-decoder architecture is handed down to explore the temporal data relationship. The model is compared with ARIMA, SVR, Bayesian connected-LSTM (B-LSTM), Diffusion convolutional RNN, and Temporal GCN and recorded significant improvement in MAE, MAPE, and RMSE for city-wide traffic speed prediction.

Similar to traffic flow prediction, traffic speed forecasting is required to analyze the congestion on-road segment in the future and provide guidance to the drivers to estimate the pass-through of the targeted road segment.

Travel Time Prediction

The required time duration in transporting of goods plays the major roles in terms of the efficient transportation management system. However, travel time forecasting is a challenging and interesting problem and is efficiently addressed using deep learning techniques. A few of the major techniques are presented in this section as follows:

Deep MLP with data preprocessing [49]. A multi-step deep learning-based algorithm starts with the data preprocessing and is passed through feature analysis, feature extraction, and clustering algorithm to improve Spatio-temporal feature space. The features are represented using a deep-stacked autoencoder and decoder with a dropout layer. Finally, the deep MLP is trained to predict the travel time. The framework of the model is depicted in Fig. 14.10. The traffic historical data is given to the initial stage for outlier fitment and then loaded for diversifying the training data, feature analysis, and extraction in the second and third stages respectively. Before feeding the data to deep MLP for prediction, it is being represented through a stacked autoencoder.

The parameters MAE, MAPE, and RMSE are validated against applying raw data directly to the deep neural network and claimed on an average betterment of 4 min. The model is robust against the overfitting problem.

Hybrid model coupling the deep learning model and the quantile regression (QR) [50]. In this study, a state-space matrix is created to correlate spatial–temporal dependencies of traffic data. Feature analysis and extraction process are accomplished

Fig. 14.10 Feature integration travel time prediction: Architecture [49]

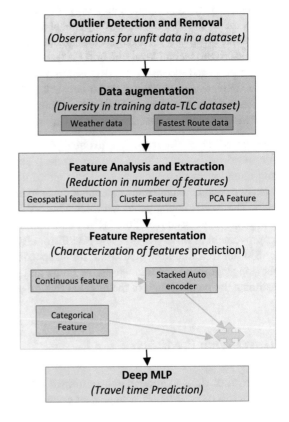

using several stacked Gaussian Bernoulli Restricted Boltzmann Machine (GBRBM). However, the final prediction is obtained through the regression layer of the deep belief neural network. The accuracy of the model is further improved with QR to generate the prediction interval. The real-time datasets are applied to the proposed model and contrasted against the benchmarked models.

The evaluation metrics, such as MAE, RMSE, and MAPE measured against the ARIMA, BPNN, LSTM, and PSO-ELM for deterministic travel time prediction. However, average coverage error (ACE), interval sharpness (IS), and continuous rank probability score (CRPS) metrics are validated for one-step and multi-step probabilistic travel time prediction for similar benchmark models and claimed better accuracy of prediction.

Travel time estimation depends on the traffic congestion, weather in a particular road segment, current road state, and so on. Deep learning models presented above are somehow able to forecast the short term in advance so that the appropriate decision can be made by the commuters.

Traffic Congestion Prediction

The number of vehicles on road is increasing day by day and thus traffic congestion management appears like a demanding research area in the domain of transportation management. Deep neural network-based solution approaches garnered much attention in the research community to provide correlation of Spatio-temporal dependencies and establish an efficient predictive performance in terms of prediction errors.

LSTM neural network-based region-wide traffic congestion model [51]. In this research work, Newell's simplified theory of kinematic waves is used to predict traffic congestion. The suitability and scalability of the model are tested for neighborhood traffic conditions. The proposed graph-CNN and LSTM neural network-based attention model bettered the accuracy of prediction against the Holt-Winters (HW) model, 1-Nearest Neighbor (1-NN) model, and Random Forest (RF) model for four types of signals that are:

(i)   Regional parameters in the context of travel requirement are categorically separated by source and destination
(ii) Average travel time of vehicles in the regional linkages are identified
(iii) Vehicle assemblage on the linkages are captured
(iv) The congestion status around the targeted space known as the congestion score is used in the research work.

All the above four signals have been experimented with for traffic congestion prediction and evaluated through RMSE and concluded better in result in terms of daily uncertainty in demand and model scalability with higher network size.

PrePCT [52]. A relative position congestion tensor is created to support traffic congestion prediction. The specific region is selected to record the traffic data and converted into three-dimensional spatial–temporal tensors and applied over LSTM NN to forecast the congestion in near future. The proposed model is compared with the baseline frameworks, such as linear regression, support vector regression, random forest, gradient boosting regression, and LSTM. The comprehensive experimentation is conducted over the PeMS dataset to compare with baseline prediction techniques in respect of MAE and RMSE.

Subsequently, the temporal traffic speed data, which is correlated with multi-dimensional context information is used to predict traffic congestion and achieved 90% accuracy [53]. Similarly, event- and time-related data and classified to predict using a deep learning network and obtained accuracy up to 98% in [54].

Traffic congestion slows down the traffic speed and flows and thus needs to be addressed carefully.

Travel Risk Prediction

To reduce the adverse effect of road accidents and crashes among the high-speed vehicle and even low-speed vehicles on congested roads is required to take appropriate prior measures by the traffic management authorities. A few major deep learning-based solutions are explored below.

Spatiotemporal Convolutional Short-term Memory Network (STCM-Net). Accident risks are increasing as the vehicles on the road are increasing. Short-term risk forecasting is presented by Bao et al. [55]. Multiple datasets including weather, population near the prediction area, crash data, large-scale taxi GPS data, road topology data, etc., are collected to propose spatiotemporal convolutional short-term memory network (STCM-Net) for risk prediction. Depending on the data collected, three types of variables are considered during experimentation:

(i)   Varying spatial data while keeping temporal as stationary especially long-term data collection.
(ii)   Temporal data are used while keeping spatial dependencies as static.
(iii) Considering both spatial and temporal data and building a correlation for prediction.

The experimented results are claimed better in the accuracy of prediction and guide the transportation system management in a better way. The deep learning-based model is validated against the various machine learning-based benchmark models, such as ARIMA, random-parameter, random-effect, and GWR model to establish the improvement in terms of MSE, MAE, and MAPE.

In addition, it is also analyzed that the proposed model performance decreases as the spatiotemporal resolution of prediction and thus, poorly performs in terms of accuracy of prediction in hourly risk forecasting.

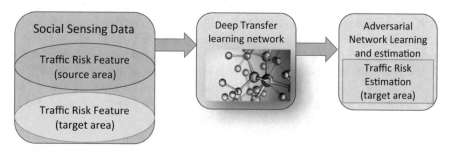

Fig. 14.11 Overview of DeepRisk

Deep learning-based CNN model with a drop-out feature. The feasibility of deep learning-based solutions is explored in [56] over the real-time collected data through the roadside radar sensors. The traffic volume, speed, and sensor occupancy data are used for the experimentation. The crash risk forecasting is estimated in a short-advance time like 1-min, 5-min, and 10-min. A deep learning-based CNN model with a drop-out feature is used to classify traffic crash occurrences and forecast crash risk.

DeepRisk [57]. Social sensing data, i.e., social media posts are used to predict road risk. The overview of the model is depicted in Fig. 14.11. Deep transfer learning network (DTNC) is used to learn the model using an extracted feature of the previous stage (first stage) and adversarial network learning and estimation (ADNL) measures the risk forecasting.

An improvement is recorded during simulation in terms of MAE over the baseline frameworks.

India reported over 151 thousand fatalities in the year 2019 in such road crashes, whereas the USA evidenced over 38 thousand lives every year. China recorded more than 60 thousand such lives every year. Statistics say 145 thousand collisions in Rushia in the year 2020. To reduce this huge fatality rate, data-driven road risk forecasting is required.

Traffic Pollution Monitoring

One of the air pollution contributors is vehicle emissions. Identifying the vehicle disobeying the rules of the pollution control board can be dealt with effectively using deep learning models using an IoT system. An image-based transfer learning model is presented by Kundu and Maulik [58]. The images are captured through another IoT-enabled vehicle or surveillance camera installed on the roads and can be fed to the monitoring control room for further actions. The data or image captured is processed through a deep learning model in real-time and work on low light and bad weather condition.

A combination of CNN and LSTM (ConvLSTM) [59] is modeled for traffic pollution prediction in Seoul city of Koria. Traffic volume and average vehicle speed is two major concern of air pollution in terms of road traffic. Spatial and temporal traffic dependencies are addressed in this research to predict the air quality of the city.

An association of RNN and LSTM is used to estimate CO_2 emission in real-time via in-vehicle sensors [60]. The sensor data captured in real-time is uploaded to the cloud and provides the result to the onboard diagnosis port. The scalability and efficiency of the model are validated over public vehicle data.

The expansion of urbanization and cutting trees are the major cause of air pollution. In addition, increasing vehicle volume is catalyzing air pollution which needs to care for upcoming generations.

Parking Occupancy Prediction

Restricted roadside parking and limited parking space availability create uttermost traffic congestion and bottlenecks at the crossing and red light signals.

Clustering Augmentation for Long short-term Memory (CALM). CNN and LSTM based deep feature learning method called CALM is proposed to solve the parking occupancy in [61]. CALM consists of simultaneous heterogeneous clustering and regression learning for spatial–temporal traffic data. The block-level parking occupancy is evaluated against the multi-layer LSTM and LASSO, obtained 7.8% mean absolute percentage error.

Block-level parking. A block-level parking occupancy is presented through GCNN for spatial data featuring and RNN with LSTM for temporal data collection [62]. Based on these two ensembles, the parking occupancy is predicted and compared with multi-layer LSTM, MSTARMA, Kalman filter, SVR, and LASSO with MAPE 10.6%. The prediction of parking occupancy is generated 10 min in advance. Weather condition and traffic speed near the block area is considered during the experimentation. The effective parking guidance to the driver and dynamic parking charges are also addressed in the research and address the enforcement parking pricing.

To avoid traffic congestion and time loss due to searching of available parking slots, simultaneous traffic and parking slot forecasting is presented in [63]. The road and the parked car is considered as a vector that inculcates both spatial and temporal information.

Vehicle parking management near market sites, over congested roads, and residential places is vital in terms of honoring neighborhoods and passengers. To predict the availability of parking slots in nearby places ease the mind tightness of drivers.

Chapter Conclusion and Future Scope

Intelligently managing transportation systems is the demand of the current era. The changing lifestyle and demolishing nature are depriving our ecosystem. Deep learning-based technologies demand a huge amount of data to learn the system and thus forecast for a better future. Capturing the data in real-time is not an easy task, but requires a sufficient amount of noise cancelation that comes up along with the required data. The noise present in the data hampers the prediction madel efficiency and aquiracy. However, IoT provides device-to-device communication where objects can share information and take decisions accordingly.

In this chapter, the transportation management issues are addressed, and artificially intelligent-based solutions are presented. Traffic flow, speed, and congestion forecasts are addressed using various multi-layer neural networks.

Extensive research has been yet conducted in the field of transportation management. However, as a future direction the research can be carried out:

(i)   The timespan to predict the traffic flow, speed, congestion in advance is quite low and needs to expand a bit more so that the drivers can plan well before the target location. However, the task is not easy due to so many spatial and temporal dependencies but can be advanced with new specialized filters and learning-data storage techniques.

(ii) The accuracy level of travel time and risk prediction is not good enough and needs to be considered with more realistic parameters, such as impulsive traffic in a road segment. The area-specific research is required rather than generalized.

(iii) Though, the research in the field of vehicle pollution detection and surveillance call has been conducted, a lot has yet to come. Such automatic pollution detection can be a game changer for developing countries like India which is suffering a lot with the balance in the ecosystem.

(iv) Another direction of work that needs to emphasis is security and safety. The cloud IoT-based solutions can ensure the safety of the vehicle to the owner.

A wide range of research articles are available but many more are required to address the practical scenarios of the transportation system. As the smart city project and industrial revolution 4.0 are underway in developing countries, the usage of technology is expected to enrich the same to some extent.

References

1. van Dongen, Leo, A.M., Frunt, L., Rajabalinejad, M.: Issues and challenges in transportation. In: Transportation Systems, pp. 3–17. Springer, Singapore (2019)
2. Tyan, J.C., Wang, F.K., Du, T.: Applying collaborative transportation management models in global third-party logistics. Int. J. Comput. Integr. Manuf. **16**(4–5), 283–291 (2003)
3. Faulin, J., Lera-López, F., Juan, A.A.: Optimizing routes with safety and environmental criteria in transportation management in Spain: a case study. Int. J. Inf. Syst. Supply Chain Manag. (IJISSCM) **4**(3), 38–59 (2011)

4. Singh, S.K.: Review of urban transportation in India. J. Public Transp. **8**(1), 5 (2005)
5. Msigwa, R.E.: Challenges facing urban transportation in Tanzania. Math. Theory Model. **3**(5), 18–26 (2013)
6. Loveless, S.M., Welch, J.S.: Growing to meet the challenges: Emerging roles for transportation management associations. Transp. Res. Rec. **1659**(1), 121–128 (1999)
7. Wang, F.-Y.: The emergence of intelligent enterprises: from CPS to CPSS. IEEE Intell. Syst. **25**(4), 85–88 (2010)
8. Stankovic, J.A.: Research directions for the Internet of Things. IEEE Internet Things J. **1**(1), 3–9 (2014)
9. Wang, F.-Y.: Scanning the issue and beyond: crowdsourcing for field transportation studies and services. IEEE Trans. Intell. Transp. Syst. **16**(1), 1–8 (2015)
10. Athreya, A.P., Tague, P.: Network self-organization in the Internet of Things. In: Proceedings of IEEE International Conference on Sensor, Communiation and Network (SECON), pp. 25–33 (2013)
11. Chen, S., Xu, H., Liu, D., Hu, B., Wang, H.: A vision of IoT: applications, challenges, and opportunities with China perspective. IEEE Internet Things J. **1**(4), 349–359 (2014)
12. Goodfellow, I., Bengio, Y., Courville, A.: Deep Learning. MIT Press (2016)
13. Kelleher, J.D.: Deep Learning. MIT press (2019)
14. Abiodun, O.I., Jantan, A., Omolara, A.E., Dada, K.V., Mohamed, N.A., Arshad, H.: State-of-the-art in artificial neural network applications: a survey. Heliyon **4**(11), e00938 (2018)
15. Zhang, Z.: Artificial neural network. In: Multivariate Time Series Analysis in Climate and Environmental Research, pp. 1–35. Springer, Cham (2018)
16. Bhaskaran, P.E., Maheswari, C., Thangavel, S., Ponnibala, M., Kalavathidevi, T., Sivakumar, N.S.: IoT Based monitoring and control of fluid transportation using machine learning. Comput. Electr. Eng. **89**, 106899 (2021)
17. Husni, E., Hertantyo, G.B., Wicaksono, D.W., Hasibuan, F.C., Rahayu, A.U., Triawan, M.A.: Applied Internet of Things (IoT): car monitoring system using IBM BlueMix. In: 2016 International Seminar on Intelligent Technology and Its Applications (ISITIA), pp. 417–422. IEEE (2016)
18. Ben-Daya, M., Hassini, E., Bahroun, Z.: Internet of things and supply chain management: a literature review. Int. J. Prod. Res. **57**(15–16), 4719–4742 (2019)
19. de Vass, T., Shee, H., Miah, S.J.: IoT in supply chain management: a narrative on retail sector sustainability. Int. J. Logistics Res. Appl. 1–20 (2020)
20. Gao, Q., Guo, S., Liu, X., Manogaran, G., Chilamkurti, N., Kadry, S.: Simulation analysis of supply chain risk management system based on IoT information platform. Enterp. Inf. Syst. **14**(9–10), 1354–1378 (2020)
21. Abdel-Basset, M., Manogaran, G., Mohamed, M.: Internet of Things (IoT) and its impact on supply chain: a framework for building smart, secure and efficient systems. Futur. Gener. Comput. Syst. **86**, 614–628 (2018)
22. Elkin, D., Vyatkin, V.: IoT in traffic management: review of existing methods of road traffic regulation. In: Computer Science On-line Conference, pp. 536–551. Springer, Cham (2020)
23. Masek, P., Masek, J., Frantik, P., Fujdiak, R., Ometov, A., Hosek, J., Andreev, S., Mlynek, P., Misurec, J.: A harmonized perspective on transportation management in smart cities: the novel IoT-driven environment for road traffic modeling. Sensors **16**(11), 1872 (2016)
24. Zhu, F., Lv, Y., Chen, Y., Wang, X., Xiong, G., Wang, F.-Y.: Parallel transportation systems: toward IoT-enabled smart urban traffic control and management. IEEE Trans. Intell. Transp. Syst. **21**(10), 4063–4071 (2019)
25. Avatefipour, O., Sadry, F.: Traffic management system using IoT technology-A comparative review. In: 2018 IEEE International Conference on Electro/Information Technology (EIT), pp. 1041–1047. IEEE (2018)
26. Mainetti, L., Patrono, L., Stefanizzi, M.L., Vergallo, R.: A Smart Parking System based on IoT protocols and emerging enabling technologies. In: 2015 IEEE 2nd World Forum on Internet of Things (WF-IoT), IEEE, pp. 764–769 (2015)

27. Khanna, A., Anand, R.: IoT based smart parking system. In: 2016 International Conference on Internet of Things and Applications (IOTA), pp. 266–270. IEEE (2016)
28. Ajchariyavanich, C., Limpisthira, T., Chanjarasvichai, N., Jareonwatanan, T., Phongphanpanya, W., Wareechuensuk, S., Srichareonkul, S., et al.: Park King: an IoT-based smart parking system. In: 2019 IEEE International Smart Cities Conference (ISC2), pp. 729–734. IEEE (2019)
29. Abdulkader, O., Bamhdi, A.M., Thayananthan, V., Jambi, K., Alrasheedi, M.: A novel and secure smart parking management system (SPMS) based on integration of WSN, RFID, and IoT. In: 2018 15th Learning and Technology Conference (L&T), pp. 102–106. IEEE (2018)
30. Ali, G., Ali, T., Irfan, M., Draz, U., Sohail, M., Glowacz, A., Sulowicz, M., Mielnik, R., Faheem, Z.B., Martis, C.: IoT based smart parking system using deep long short memory network. Electronics 9(10), 1696 (2020)
31. Bengio, Y., Goodfellow, I., Courville, A.: Deep Learning, vol. 1. MIT press, Massachusetts, USA (2017)
32. Guo, Y., Liu, Y., Oerlemans, A., Lao, S., Wu, S., Lew, M.S.: Deep learning for visual understanding: a review. Neurocomputing 187, 27–48 (2016)
33. Schmidhuber, J.: Deep learning in neural networks: an overview. Neural Netw. 61, 85–117 (2015)
34. Wang, Y., Zhang, D., Liu, Y., Dai, B., Lee, L.H.: Enhancing transportation systems via deep learning: a survey. Transportation Research Part C: Emerging Technologies 99, 144–163 (2019)
35. Kong, F., Li, J., Jiang, B., Song, H.: Short-term traffic flow prediction in smart multimedia system for Internet of Vehicles based on deep belief network. Futur. Gener. Comput. Syst. 93, 460–472 (2019)
36. Dai, X., Fu, R., Zhao, E., Zhang, Z., Lin, Y., Wang, F.-Y., Li, L.: DeepTrend 2.0: a light-weighted multi-scale traffic prediction model using detrending. Transp. Res. Part C Emerg. Technol. 103, 142–157 (2019)
37. Li, Z., Li, Y., Li, L.: A comparison of detrending models and multi-regime models for traffic flow prediction. IEEE Intell. Transp. Syst. Mag. 6(4), 34–44 (2014)
38. Polson, N.G., Sokolov, V.O.: Deep learning for short-term traffic flow prediction. Transp. Res. Part C Emerg. Technol. 79, 1–17 (2017)
39. Wu, Y., Tan, H., Qin, L., Ran, B., Jiang, Z.: A hybrid deep learning based traffic flow prediction method and its understanding. Transp. Res. Part C Emerg. Technol. 90, 166–180 (2018)
40. Cho, K., Van Merriënboer, B., Gulcehre, C., Bahdanau, D., Bougares, F., Schwenk, H., Bengio, Y.: Learning Phrase Representations Using RNN Encoder-Decoder for Statistical Machine Translation. arXiv preprint arXiv: 1406.1078. (2014)
41. Sutskever, I., Vinyals, O., Le, Q.V.: Sequence to sequence learning with neural networks. In: Advances in Neural Information Processing Systems, pp. 3104–3112 (2014)
42. Hao, S., Lee, D.-H., Zhao, D.: Sequence to sequence learning with attention mechanism for short-term passenger flow prediction in large-scale metro system. Transp. Res. Part C Emerg. Technol. 107, 287–300 (2019)
43. Miglani, A., Kumar, N.: Deep learning models for traffic flow prediction in autonomous vehicles: a review, solutions, and challenges. Veh. Commun. 20, 100184 (2019)
44. Wang, J., Chen, R., He, Z.: Traffic speed prediction for urban transportation network: a path based deep learning approach. Transp. Res. Part C Emerg. Technol. 100, 372–385 (2019)
45. Zhang, Z., Li, M., Lin, X., Wang, Y., He, F.: Multistep speed prediction on traffic networks: a deep learning approach considering spatio-temporal dependencies. Transp. Res. Part C Emerg. Technol. 105, 297–322 (2019)
46. Gu, Y., Wenqi, L., Qin, L., Li, M., Shao, Z.: Short-term prediction of lane-level traffic speeds: a fusion deep learning model. Transp. Res. Part C Emerg. Technol. 106, 1–16 (2019)
47. Zhang, K., Zheng, L., Liu, Z., Jia, N.: A deep learning based multitask model for network-wide traffic speed prediction. Neurocomputing 396, 438–450 (2020)
48. James, J.Q.: Citywide traffic speed prediction: a geometric deep learning approach. Knowl.-Based Syst. 212, 106592 (2021)
49. Abdollahi, M., Khaleghi, T., Yang, K.: An integrated feature learning approach using deep learning for travel time prediction. Expert Syst. Appl. 139, 112864 (2020)

50. Li, L., Ran, B., Zhu, J., Bowen, D.: Coupled application of deep learning model and quantile regression for travel time and its interval estimation using data in different dimensions. Appl. Soft Comput. **93**, 106387 (2020)

51. Mohanty, S., Pozdnukhov, A., Cassidy, M.: Region-wide congestion prediction and control using deep learning. Transp. Res. Part C Emerg. Technol. **116**, 102624 (2020)

52. Bai, M., Lin, Y., Ma, M., Wang, P., Duan, L.: PrePCT: traffic congestion prediction in smart cities with relative position congestion tensor. Neurocomputing **444**, 147–157 (2021)

53. Kim, D.H., Hwang, K.Y., Yoon, Y.: Prediction of traffic congestion in seoul by deep neural network. J. Korea Inst. Intell. Transp. Syst. **18**(4), 44–57 (2019)

54. Sun, F., Dubey, A., White, J.: DxNAT—Deep neural networks for explaining non-recurring traffic congestion. In: 2017 IEEE International Conference on Big Data (Big Data), pp. 2141–2150. IEEE (2017)

55. Bao, J., Liu, P., Ukkusuri, S.V.: A spatiotemporal deep learning approach for citywide short-term crash risk prediction with multi-source data. Accid. Anal. Prev. **122**, 239–254 (2019)

56. Huang, T., Wang, S., Sharma, A.: Highway crash detection and risk estimation using deep learning. Accid. Anal. Prev. **135**, 105392 (2020)

57. Zhang, Y., Wang, H., Zhang, D., Wang, D.: Deeprisk: A deep transfer learning approach to migratable traffic risk estimation in intelligent transportation using social sensing. In: 2019 15th International Conference on Distributed Computing in Sensor Systems (DCOSS), pp. 123–130. IEEE (2019)

58. Kundu, S., Maulik, U.: Vehicle pollution detection from images using deep learning. In: Intelligence Enabled Research, pp. 1–5. Springer, Singapore (2020)

59. Le, V.-D., Bui, T.-C., Cha, S.-K.: Spatiotemporal deep learning model for citywide air pollution interpolation and prediction. In: 2020 IEEE International Conference on Big Data and Smart Computing (BigComp), pp. 55–62. IEEE (2020)

60. Ghosal, S.S., Bani, A., Amrouss, A., El Hallaoui, I.: A deep learning approach to predict parking occupancy using cluster augmented learning method. In: 2019 International Conference on Data Mining Workshops (ICDMW), pp. 581–586. IEEE (2019)

61. Yang, S., Ma, W., Pi, X., Qian, S.: A deep learning approach to real-time parking occupancy prediction in transportation networks incorporating multiple spatio-temporal data sources. Transp. Res. Part C Emerg. Technol. **107**, 248–265 (2019)

62. Ziat, A., Leroy, B., Baskiotis, N., Denoyer, L.: Joint prediction of road-traffic and parking occupancy over a city with representation learning. In: 2016 IEEE 19th International Conference on Intelligent Transportation Systems (ITSC), pp. 725–730. IEEE (2016)

63. Lu, E.H.C., Liao, C.H.: A parking occupancy prediction approach based on spatial and temporal analysis. In: Asian Conference on Intelligent Information and Database Systems, pp. 500–509. Springer, Cham (2018)

Chapter 15
Enhancing Shoppers' Loyalty by Prioritizing Customer-Centricity Drivers in the Retail Industry

Vishal Srivastava, Manoj Kumar Srivastava, and R. K. Singhal

Introduction

With the emergence of digital technology, the gap between shoppers and sellers has minimized. Moreover, the change in global economic policies has also minimized the gap between physical retail-stores and buyers. Globalization of retailing initiated the establishment of a mature market retailer into the other various economic stage markets, globalizing the supply chain and logistics and innovations in retailing practices [40]. No doubt, these changes are helping the retail industry, but in another way, the retailers are facing a lot of challenges, like competition, supply chain, and retail-knockoffs [41] and others, in retaining their customer base. Following the traditional marketing concept, firms believe in delivering higher customer satisfaction to win the market [3, 35]. In practice, better quality services delivery has proven in strong association with better customer retention [51]. To achieve higher customer satisfaction and protect their market share, retailers are applying various practices. Enabling customer-centricity is one such approach, which is advocated by the number of marketing philosophers [37]. As an approach, customer centricity is considered as the focus level of organization on understanding customers need and offering customer-focused solutions [16]. However, offering the optimum number of desired benefits or services to the customer is always invited by the market [30]. Customer centricity means a change in organization focus toward customer profitability, customer lifetime value, customer equity, and from market share to customer equity share [42]. The advanced tools of business intelligence can also assist in enabling the customer-centric approach [20]. Overloading the undesired

V. Srivastava (✉) · R. K. Singhal
Dr A P J Abdul Kalam Technical University, Lucknow, India
e-mail: drvsmba@gmail.com

M. K. Srivastava
Management Development Institute, Gurgaon, India

services or offerings is not only ineffective but also demotivate customers [27]. Referring to the basic concept of marketing, the philosophers are raising the question toward a customer-centric approach and defining it as "paradigmatic trap" in securing the organizations' commercial existence [37]. Thus, enabling customer centricity must be done carefully to secure the well-beingness of both, the customer and the organization.

In this article, we identified the key drivers to establish customer-centricity specifically for retail operations in stores. The study is based on the various-related studies and focuses on customer opinion toward the suggested customer-centricity drivers.

Customer Centricity Drivers

The evolution of digital technology enabled the ease in assessing marketer and consumer to each other. The parallel evolved data science tools can easily understand the complex buying behavior of customers, which was not as easy earlier [33]. To implement customer centricity, organizations need to understand the desired services or offers or factors by the customers. It is observed that many organizations only claim for customer centricity, but not in their practices [47] due to a lack of clarity about customer need and desires. Some studies classified the customer-centricity driving variables into three categories such as environmental level, organizational level, and department level [55].

To identify the drivers of customer-centricity, the study surveyed various concomitant literature. The study undertook references from various industries of various periods. The study identified unique 92 customer-centricity driving factors/drivers from the intensive literature survey. The details are given in Table 15.1. As discussed above, such large variables create a status of confusion. The marketers find themselves in an indecisive position in framing the policy to adopt a customer-centric approach. Addition or removal of any operational activity or process is always a costlier affair. On the name of customer-centricity, one can blindly add or remove any feature or activity in his or her commercial system.

The customer-centricity is defined as the orientation toward the well-being of customers [1] through fulfilling their needs and offering required services. The studies also advocate that service differentiation gives a competitive advantage to marketers [19]. But offering all desired services or fulfilling the need of customer is not always profitable for the organization. Organizations have to trade-off between customer centricity and organization centricity.

This study addresses major challenges to trade-o among various suggested driving variables in attaining customer-centricity specifically in the retail industry. As it is very much essential to minimize the cost to enable the customer-centricity to maximize the customer need fulfillment.

Table 15.1 Customer centricity driving variables

S.No.	Reference	Variable
1	[47]	Managing relationship with customer, highlight produc's benefits
2	[2]	Retail store devoted to customer-convenience
3	[4]	Influencing colleagues, friend, and relatives
4	[5]	Competent counter staff
5	[6]	Proper communication with customer
6	[7]	Less time-consumption in shopping process
7	[9]	Offering credit services
8	[10]	Zero defect in first time
9	[13]	Quality information sharing processes
10	[14]	Quality of products, number of own brands, well-known brands, quality of fresh counters, high frequency of discount days, close to home, close to work, effective traffic and parking management, impressive layout design, much spacious to ease of driving trolleys, ease of shopping, express checkouts facilities, exchange guarantee policies, cash back offer, servicing customer type
11	[15]	Customer-focused innovation, seamless network operations, opportunity-creating partnerships, value-adding services, specificity of products and services
12	[17]	Good relationships with key-suppliers, proper detailing the customization-desires, ensuring value for customer
13	[18]	Conference to requirement
14	[23]	RFID-enabled payment-system
15	[25]	Value creation processes
16	[31]	Sufficient parking lot, offering baby areas, delivery services of goods, freshness of products, offering durable product, high product variety, soothing music and ambience, correct price practices, high product variety, offering competitive price, offering membership card
17	[32]	Low-price offerings
18	[36]	Innovative membership programs
19	[38]	Huge product assortment
20	[39]	Accuracy in billing, providing transactional document immediately, quick customer call-back, competent store personnel, ensuring financial security, assure confidentiality of shared-information, ease to reach the store, proper explanation of services, assure Confidentiality of shared-information, use of suitable communication-media, proper documentation of records, impressive ambience of store, service performance in a responsible manner, eagerness of service personnel, enthusiasm of service personnel, up-to-date equipment, visually appealing physical-facilities, well appeared employee, proper documentation of commitments, timely fulfilling promised services, sympathetic toward customer problems, trustworthiness of retail store, timely delivery the services, keep records accurately, sympathetic toward customer problems, employee polite with customer, employees get proper support from store, store give individual-attention to customers, operating-hours convenient to customers

(continued)

Table 15.1 (continued)

S.No.	Reference	Variable
21	[44]	Providing services online
22	[45]	Convenience in product search, convenience in product possession, convenient telephone and Internet access, intelligent customer-interactive systems, strong in-stock position, one-stop shopping offer
23	[48]	Unique product assortment, inimitable product assortment, establishing strong value-proposition
24	[49]	Continuous after-sale services
25	[53]	Friendly sales persons
26	[54]	Bundling good and services

Problem Discussion

From early 1950s, various management and marketing Gurus are advocating customer-first mantra, as a key to success [12, 29]. In Gartner Group Report-2003, not only emphasized on customer-centric approach but also advocated it as the best tool for success in coming time [33]. Globally, every organization is convinced with the significance of the customer-centric approach as a strategic move for growth and survival [11]. Despite this, even after six decades of emergence, large number of organizations are trying hard and facing challenges to become customer-centric in true meaning [33].

When we talk about the retail industry, the situation becomes more challenging. With globalization and paced technology upgrades, the retail industry is also changing parallelly. The concept of customer-centricity seems easy to adopt but imposes a challenge to establishment [21]. The challenges to establish customer-centricity are internal as well as external. Among various challenges to establish customer-centricity, non-clarity about the need and desires of customers is a major one. There is no shortcut to establish customer-centricity [22].

The purpose of this study is to identify the most significant customer-centricity enabling drivers which are appreciated by the shoppers. The study intends to develop a model for retailers, which can help them to decide a more systematic and structured course of actions to adopt the customer-centric approach.

Research Methodology

This exploratory and descriptive study are conducted to know about the most apprenticing customer-centricity driving factors toward the retail industry. Initially, 92 such variables identified from extensive literature survey (Table 15.1). Many of the identified variables are found related or like each other. To reduce the number of variables and make them more applicable, the variables/drivers are sorted, using the text selection sort technique (TST) [46]. The TST is a proven tool to group the resembling variables [52]. The sorting reduced 92 identified variables into twenty drivers (Table 15.2).

To understand the customer need, a survey was applied to a sample of 1200 shoppers to assess buyer preferences and expectations regarding retailing operations. The study applied sampling random sampling to select the sample from an urban

Table 15.2 Customer centricity drivers

S.No.	Driver	Variables (TLA)
1	Ambiance (D1)	Impressive layout design (ILD) Soothing music and ambience (SMA) Impressive ambience of store (IAS) Visually appealing physical-facilities (VAP)
2	Care to customer (D2)	Assure confidentiality of shared information (ACS) Service performance in a responsible manner (SRM) Managing relationship with customer (MRC) Quick customer call-back (QCC) Eagerness of service personnel (ESP) Sympathetic toward customer problems (SCP) Trustworthiness of retail store (TRS) Store give individual-attention to customers (SIC) Friendly sales persons (FSP)
3	Convenience (D3)	Retail store devoted to customer-convenience (RDC) Closeness to home (CtH) Closeness to work (CtW) Much spacious to ease of driving trolleys (EDT) Ease of shopping (EoS) Ease to reach the store (ERS) Operating-hours convenient to customers (OCC) Providing services online (PSO) Convenient telephone and internet access (CTI) One-stop shopping offer (OSO)
4	Delivery services (D4)	Delivery services of goods (DSG) Good number of checkout counters (GCC) Express checkouts facilities (ECF) Timely delivery the services (TDS)

(continued)

Table 15.2 (continued)

S.No.	Driver	Variables (TLA)
5	Information sharing (D5)	Use of suitable communication-media (USC) Highlight product's benefits (HPF) Proper detailing the customization desires (PDC) Quality information sharing processes (QIS) Proper explanation of services (PES) Proper communication with customer (PCC)
6	Payment service (D6)	Offering credit services (OCS) RFID-enabled payment-system (REP) Competent counter staff (CCS) Accuracy in billing (AIB)
7	Post purchase services (D7)	Conference to requirement (CTR) Exchange guarantee policies (EGP) Opportunity-creating partnerships (OCP) Continuous after-sale services (CAS)
8	Pricing policy (D8)	Correct price practices (CPP) Proper explanation of services-cost (PEC) Offering competitive price (OCP) Low-price offerings (LPO)
9	Product availability (D9)	Good relationships with key-suppliers (GRK) Strong in-stock position (SIP)
10	Product quality (D10)	Quality of products (QoP) Well-known brands (WKB) Quality of fresh counters (QFC) Freshness of products (FoP) Offering durable product (ODP)
11	Product variety (D11)	Number of own brands (NOB) High product variety (HPV) Huge product assortment (SoA)
12	Promotion policy (D12)	High frequency of discount days (HFD) Cash back offer (CBO) Innovative membership programs (IMP) Offering membership card (OMC)
13	Proper documentation/transparency (D13)	Providing transnational document immoderately (PTD) Ensuring financial security (EFS) Proper documentation of records (PDR) Keep records accurately (KRA) Proper documentation of commitments (PDC)

(continued)

Table 15.2 (continued)

S.No.	Driver	Variables (TLA)
14	Shopping assistance services (D14)	Less time-consumption in shopping process (LTS) Bundling good and services (BGS)
15	Social image (D15)	Servicing customer type (SCT) Influencing colleagues, friend, and relatives (ICFR)
16	Staff effectiveness (D16)	Well appeared employee (WAE) Convenience in product search (CPS) Convenience in product possession (CPP) Zero defect in first time (ZDF) Competent store personnel (CSP) Employees get proper support from store (EPS) Enthusiasm of service personnel (EnSP) Timely fulfilling promised services (TFP) Employee polite with customer (EPC)
17	Advance technology (D17)	Seamless network operations (SNO) Up-to-date equipment (UDE) Customer-focused innovation (CFI) Intelligent customer-interactive systems (ICS)
18	Uniqueness (D18)	Specificity of products and services (SPS) Unique product assortment (UPA) Inimitable product assortment (IPA) Product comb packs (PCP)
19	Value added services (D19)	Value-adding services (VAS) Effective traffic and parking management (ETM) Sufficient parking lot (SPL) Offering baby areas (OBA)
20	Value to visit (D20)	Value creation processes (VCP) Ensuring value for customer (EVC) Establishing strong value-proposition (ESV)

and semi-urban population from pan India. Among the identified twenty drivers, we tried to identify more significant drivers which retailers need to fulfill to attain customer-centricity. A structured questionnaire is used to collect the data. The tool is composed of 05-point scale questions. The scale questions are calibrated in between least preferred to most preferred. Initially, the pilot test was conducted on 26 shoppers of Delhi-NCR location. The reliability test accepts the construct (Table 15.3: Cronbach's Alpha = 0.897).

For testing construct validity Pearson product-moment correlation to test the was used [24]. After the test, two driver variable D4 (delivery service) and variable D15

Table 15.3 Reliability statistics

	Cronbach's alpha	Alpha based on standardized items	N of items
	0.897	0.898	20

(social image) are dropped for further study. The mentioned variables shown non-significant relation with other test variables (Table 15.4).

The final construct was used for data collection from earlier mentioned 1200 shoppers. Among collected data set, 166 records were not found suitable due to some data collection errors, like an incomplete response, artificial responses, etc. Thus finally, 1034 data records are used in this study. The descriptive of the sample is given (Table 15.5).

Table 15.4 Product-moment correlation to test the validity

Driver	D1	D2	D3	D4	D5	D6	D7	D8	D9	D10
D1	1	0.735**	0.319**	0.613**	0.587**	0.681**	0.509**	0.564**	0.123**	0.427**
D2		1	0.268**	0.730**	0.570**	0.779**	0.533**	0.603**	0.191**	0.594**
D3			1	0.134**	0.091**	0.112**	0.142**	-0.107**	-0.235**	-0.083**
D4				1	0.452**	0.709**	0.488**	0.681**	0.505**	0.708**
D5					1	0.580**	0.576**	0.569**	0.169**	0.430**
D6						1	0.535**	0.693**	0.220**	0.578**
D7							1	0.558**	0.244**	0.419**
D8								1	0.534**	0.670**
D9									1	0.674**
D10										1

Driver	D11	D12	D13	D14	D15	D16	D17	D18	D19	D20
D1	0.120**	0.429**	0.652**	0.220**	0.079*	0.320**	0.437**	0.714**	0.572**	0.447**
D2	0.077*	0.173**	0.558**	-0.03	-0.04	0.112**	0.443**	0.765**	0.474**	0.525**
D3	0.445**	0.427**	0.550**	0.033	0.190**	0.211**	0.247**	0.280**	0.355**	-0.100**
D4	0.175**	0.070*	0.363**	-0.03	0.02	0.131**	0.151**	0.632**	0.405**	0.714**
D5	0.120**	0.188**	0.365**	0.298**	0.311**	0.203**	0.263**	0.743**	0.616**	0.345**
D6	-0.122**	0.189**	0.505**	0.033	-0.142**	0.113**	0.483**	0.723**	0.404**	0.572**
D7	0.081**	0.346**	0.416**	0.394**	0.181**	0.343**	0.246**	0.521**	0.588**	0.448**
D8	-0.062*	0.090**	0.251**	0.175**	0	0.185**	0.161**	0.557**	0.371**	0.719**
D9	0.182**	-0.190**	-0.227**	0.048	0.252**	0.076*	-0.460**	0.098**	0.111**	0.693**
D10	0.107**	-0.150**	0.112**	-0.067*	0.105**	0.005	-0.05	0.492**	0.273**	0.639**
D11	1	0.190**	0.188**	0.046	0.584**	0.198**	-0.224**	0.154**	0.346**	0.01
D12		1	0.655**	0.501**	0.143**	0.637**	0.338**	0.170**	0.507**	0.01
D13			1	0.232**	-0.01	0.400**	0.603**	0.548**	0.541**	0.108**
D14					0.312**	0.577**	0.02	0.070*	0.479**	0.082**
D15					1	0.238**	-0.404**	0.131**	0.443**	0
D16						1	0.06	0.114**	0.462**	0.141**
D17							1	0.423**	0.186**	-0.04
D18								1	0.595**	0.396**
D19									1	0.271**
D20										1

** Correlation is significant at the 0.01 level (2-tailed)
* Correlation is significant at the 0.05 level (2-tailed)

Table 15.5 Descriptions of sample

Descriptive	Gender	Location	Income (monthly)	Occupation	Age
Median	2.00	1.00	2.00	3.00	3.00
Mode	2 (male)	1 (metro city)	2 (from INR 25,001 to INR 50,000/-)	2 (govt/pvt job)	3 (35 years to 50 years)
Minimum	1.00	1.00	1.00	1.00	1.00
Maximum	2.00	4.00	5.00	4.00	5.00
Count	1034	1034	1034	1034	1034

For data analysis purpose, descriptive statistical methods are used in this study. To calculate the sample descriptive and variance analysis, SPSS-20 software is used. Although many software tools are available, like expert choice, super decisions, and decision lens; in this study, MS-Excel spreadsheet is used for AHP analysis.

Data Analysis

Analytic hierarchy process tools developed by Santy [43]. As discussed above, the study used the analytic hierarchy process (AHP) tool, to get a comprehensive and rational framework of the significant factors identify. The tool is in use worldwide for various decision situations [34] and identifies human judgment process [28]. It is known for ranking the identified alternatives to generate decision-making matrices [8]. The analysis uses a paired comparison square matrix of the variables under study. For pairwise comparison, the matrix used in the study is as follows

$$
\begin{bmatrix}
x_{11} & x_{12} & \cdots & x_{1n} \\
x_{21} & x_{22} & \cdots & x_{2n} \\
\vdots & \vdots & \vdots & \vdots \\
x_{1n} & x_{1n} & \cdots & x_{mn}
\end{bmatrix}
$$

Here, n and m are column and row number, respectively.

To apply AHP, the preference data were collected randomly from 142 shoppers from the urban area. As per the model of AHP, the respondents were asked to give their preferences on 09 points paired comparison scale. Statisticians and researched advocate a small number of key variables for the evaluation method of AHP. To reduce the variables (drivers), ranks in one-criterion variance analysis (ROVA) [26] are applied in the study. The method is used to identify the more consistent variables. The method ranks variables are based on their variance. The concept assumes lesser the variance, higher the rank. For AHP method with an expert survey, the study selected the top 9 variables in terms of their consistency (Table 15.6).

Table 15.6 Variable selection through one-criterion variance analysis

Variable	Variance	Consistency
D01	0.32	5[a]
D02	0.58	17
D03	0.22	3[a]
D04	0.42	12
D05	0.48	13
D06	0.33	6[a]
D07	0.2	2[a]
D08	0.4	11
D09	0.76	18
D10	0.36	8[a]
D11	0.51	15
D12	0.48	14
D13	0.39	10
D14	0.19	1[a]
D17	0.3	4[a]
D18	0.53	16
D19	0.37	9[a]
D20	0.35	7[a]

[a]Selected variables

After finalization the variables for the AHP, the study developed the *paired comparison matrix level 1*. In this table, the weights for every cell are calculated. The non-favored cell values are calculated as the reciprocal of the corresponding favorable cell value. It is denoted as (Eq. 15.1)

$$x_{mn} = \frac{1}{x_{nm}} \quad (15.1)$$

The mean scores of the paired comparison are taken to calculate the weight for the preferred variable. To calculate the weight of preference or non-preference, the study followed the below given rules.

(a) Mean score 1 to 3 = Not preferred over other
(b) Mean score 4 and 5 = Variables are indifferent
(c) Mean score 6 to 9 = Preference weights are 1, 2, 3, and 4 for score 6, 7, 8, and 9, respectively.

Thus, by calculating preference and non-preference weights, the study developed its first AHP table—"paired comparison matrix level 1 for drivers" (Table 15.7). In the process of normalizing the priority, the study followed two steps. First, every driver value is divided by the column sum (Eq. 15.2).

$$x_{nm} = \frac{x_{nm}}{\sum_{i=1}^{n} x_{nm}} \quad (15.2)$$

The newly formed matrix of output is termed as *paired comparison matrix level 2 for priority vector* (Table 15.8). Two columns are added in Table 15.8. The column sum is the column of row total after the first step of normalization.

The last column is of the second and final step of priorities weight normalization. In this step, each row sum is divided by the number of variables (nine in this study). The formula (Eq. 15.3) is

$$w_n = \frac{\sum_{i=1}^{n} x_{nm}}{n} \quad (15.3)$$

Table 15.7 Paired comparison matrix level 1 for drivers

Driver preference mean value

Variable	D01	D03	D10	D11	D14	D16	D17	D19	D20
D01	0.000	3.000	1.000	3.000	3.000	3.000	3.000	2.000	2.000
D03	0.333	0.000	0.333	0.333	0.500	2.000	2.000	2.000	2.000
D10	1.000	3.000	0.000	3.000	3.000	3.000	3.000	3.000	3.000
D11	0.333	3.000	0.333	0.000	2.000	3.000	3.000	3.000	3.000
D14	0.333	2.000	0.333	0.500	0.000	1.000	0.500	2.000	1.000
D16	0.333	0.500	0.333	0.333	1.000	0.000	0.500	0.500	0.333
D17	0.333	0.500	0.333	0.333	2.000	2.000	0.000	0.500	0.500
D19	0.500	0.500	0.333	0.333	0.500	2.000	2.000	0.000	2.000
D20	0.500	0.500	0.333	0.333	1.000	3.000	2.000	0.500	0.000
Sum	3.667	13.000	3.333	8.167	13.000	19.000	16.000	13.500	13.833

Table 15.8 Paired comparison matrix level 2 for priority vector

Variable	D01	D03	D10	D11	D14	D16	D17	D19	D20	Sum	Normalization
D01	0.091	0.039	0.100	0.041	0.077	0.000	0.031	0.037	0.024	0.440	0.049
D03	0.136	0.039	0.100	0.041	0.077	0.158	0.125	0.037	0.000	0.713	0.079
D10	0.091	0.000	0.100	0.041	0.039	0.105	0.125	0.148	0.145	0.793	0.088
D11	0.091	0.039	0.100	0.041	0.154	0.105	0.000	0.037	0.036	0.603	0.067
D14	0.136	0.039	0.100	0.041	0.039	0.105	0.125	0.000	0.145	0.729	0.081
D16	0.091	0.154	0.100	0.061	0.000	0.053	0.031	0.148	0.072	0.710	0.079
D17	0.091	0.231	0.100	0.000	0.154	0.158	0.188	0.222	0.217	1.360	0.151
D19	0.273	0.231	0.000	0.367	0.231	0.158	0.188	0.222	0.217	1.886	0.210
D20	0.000	0.231	0.300	0.367	0.231	0.158	0.188	0.148	0.145	1.767	0.196
Sum	1.000	1.000	1.000	1.000	1.000	1.000	1.000	1.000	1.000	9.000	1.000

The normalization values are established the final rating of priorities toward variables in the study. However, as the AHP tool recommends, check the consistency is required before any declaration. Santy [43] recommended the testing of the principal eigenvalue for testing the consistency. There is a relationship between normalized priority weight and summation of variables of the priority matrix. The formula Eq. (15.4) is used to calculate principal eigenvalue (λ_{max})

$$\lambda_{max} = \sum (\text{Column Rating} * \text{Row Normalize Value}) \qquad (15.4)$$

The following principal eigenvalues are found for the variables under study (Table 15.9).

In AHP, principal eigenvalue is a very important validating parameter to calculate the consistency ratio (CR) of the estimated variable. To test the consistency, [43] proposed random consistency index (RI) (Table 15.10), equation to calculate model consistency index (Eq. 15.5) and model consistency ratio (Eq. 15.6).

$$CI = \frac{\lambda_{max} - n}{n - 1} \qquad (15.5)$$

$$CR = \frac{CI}{RI} \qquad (15.6)$$

The CI for the model under study is found 0.1215 (Eq. 15.7). Whereas, the consistency ratio ($n = 9$ and RI $= 1.41$) is 0.0861 (Eq. 15.8), which is much lesser than the standard value (0.1). The calculations establish the consistency and acceptance of the model.

$$CI = \frac{\lambda_{max} - n}{n - 1} = \frac{9.923 - 9}{9 - 1} = 0.1215 \qquad (15.7)$$

$$CR = \frac{CI}{RI} = \frac{0.1215}{1.41} = 0.0861 \qquad (15.8)$$

However, after the entire analysis, considering the normalized priority weight, the observed ranks of drives are as follows (Table 15.11).

Table 15.9 Principal Eigen value (λ_{max})

Variable	D01	D03	D10	D11	D14	D16	D17	D19	D20	λ_{max}
Sum of column rating	3.6667	13	3.3333	8.1667	13	19	16	13.5	13.8333	
Normalization	0.4583	0.0881	0.2096	0.1511	0.0789	0.0488	0.0669	0.081	0.0792	
Eigen value	1.6806	1.1457	0.6986	1.2341	1.026	0.9278	1.0711	1.0934	1.0951	9.97232

Table 15.10 Random inconsistency indices (RI) For $N = 10$

N	1	2	3	4	5	6	7	8	9	10
RI	0	0	0.58	0.9	1.12	1.24	1.32	1.41	1.46	1.49

Source Satty 43

Table 15.11 Ranking of drivers in terms of priority weight

Variable	D10	D01	D11	D3	D19	D20	D14	D17	D16
Priority weight score	0.2096	0.196	0.1511	0.0881	0.081	0.0792	0.079	0.0669	0.0488
Rank	1	2	3	4	5	6	7	8	9

The Model

When we arrange the drivers in identified order of the customer preference, the outcome shows a specific type of pattern on which these drivers can be specified grouped (Fig. 15.1). These groups are as follows.

1. **Core or Essential Drivers**:
   The study identified that the top three drivers which are associated with basic shopping need of the customers. It includes product quality, product variety, and ambiance of retail store. Therefore, we can term them core or essential drivers.
2. **Experience Influencing Drivers**:

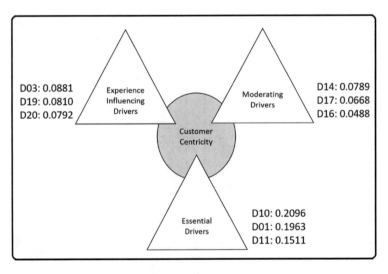

Fig. 15.1 Retail customer centricity drivers

After core drivers, the next three drivers are convenience, value added services, and value to visit. These drivers are associated with the experience of customers are not at all associated with the core services of retailing.

3. **Moderating Drivers**:

The third segment of drivers is assistance in shopping, advanced technology, and effective staff. These drives have no value in the absence of essential drivers. The presence of these drivers builds a good image of the store. The components and criteria of these drivers identified in this study are demonstrated in Fig. 15.2.

The study suggests that all three are responsible to establish customer-centricity. The hierarchical structure of these drivers and their criteria can be followed to enable customer centricity in retailing a (Fig. 15.3) model to establish customer centricity in retail per this study. Although all these three variables are required to enable customer-centricity, retailers must follow a systematic way in applying the above-mentioned three factors.

As per this model, first, retailers must ensure the availability of a large variety of good quality product in quality ambience. After essential drivers, retailers need to work upon experience influencing drivers. Finally, the presence of moderating drivers establishes the proper customer-centricity in retail operations (Fig. 15.3).

Research Implication

This study answered the question to enable customer-centricity. As discussed earlier, retailers face a lot of challenges to enable customer centricity in their operations. The basic question which remains in their mind is how to adopt customer-centricity in their retail operation?

This study provides a structured systematic way to enable the customer centricity in retailing. As a process of customer-centricity approach adoption, retailers primarily need to focus on the essential drivers along with the experience influencing drivers. Synergistic effect of these drivers is more powerful to ensure customer-centricity. Moderation drivers are ever-changing; therefore, the retailers need to upgrade their operation in a routine way (Fig. 15.3).

This study is equally applicable for organized, unorganized, and online retails. In any format of retail, the two components of the essential driver-product quality and product variety are uniformly applicable, whereas the ambience can be considered as a tangible dimension of RATER tool [50]. Therefore, for a physical store, the ambience means the look, layout, and decorations of the store, whereas for online stores, the dimension ambience refers to the Web designing, Web pages structure and content, etc. Ignorance of these prime features can never establish customer-centricity in retail operations.

To improve experience influences drivers, the retailers need to work on convenience during the shopping time, adding value to visit, delivering the assistance service to make their visit memorable. It is also equally applicable to all types of

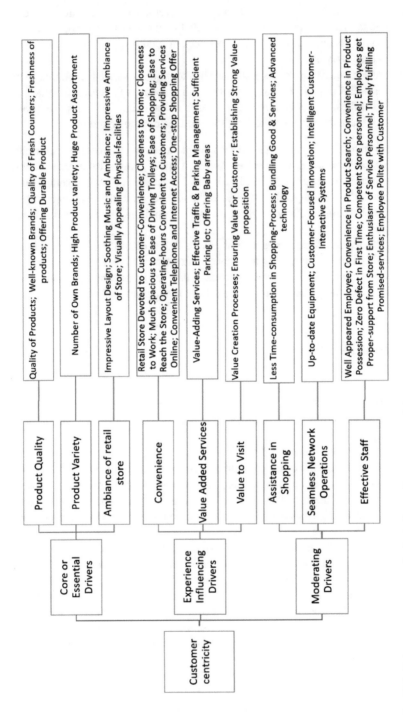

Fig. 15.2 Driver, components, and criteria for retail customer centricity

Fig. 15.3 Model to establish customer centricity in retail

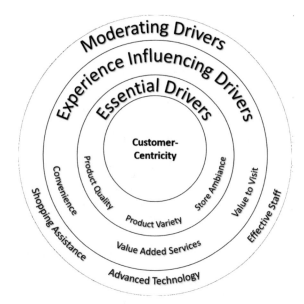

retail formats. For online retails, chat assistance, recommendation tools, the previous search supports, etc., can be used as the experience of influencing drivers.

The moderating drivers vary with retail formats. However, the basic intentions behind them remain the same. The assistance in decision-making, connectivity with customer need, and CRM must be the basic objective of any such activity, used by the retailer, as the moderating driver. In this highly competitive and dynamic environment, this model emerged as a strategical tool, retailers can count on it. However, their success will vary on the level of execution of this model.

Conclusion

The main purpose of this study is to identify the retail customer-centricity driver in align with customer preferences. This article analyzes various customer-centricity drivers in retail which affect the decision-making of retail selection. The study proposes an AHP model for retailers to enable customer centricity in their retail operations. The study uses AHP to identify the hierarchy of customer-centricity drivers. The above discussion and analysis laid and established a model which describes the customer-centricity drivers for any retail organization, in align with the priorities of the customers. This study is conducted mainly in metro and urban areas. Further studies can be conducted by including uncovered geographies sub-urban or rural areas. Moreover, this study can be extended in the focus of online retail also or by identifying new variables through an extended literature survey.

References

1. Ambaram, M., et al.: The factors that enable customer centricity and the changes in the organisation design when moving from a product to a customer centric strategy. Ph.D. Thesis, University of Pretoria (2013)
2. Anderson, W.T.: Convenience orientation and consumption behavior. J. Retail. **48**(3), 49 (1972)
3. Armstrong, Gary Agnihotri, P.K.P.T.: Principles of Marketing, 17th edn. Pearson India (2018)
4. Ashokan, C., Hariharan, G.: Profile and perception of retail consumers-an empirical study in palakkad district. Indian J. Market. **38**(2) (2008)
5. Baker, J., Parasuraman, A., Grewal, D., Voss, G.B.: The influence of multiple store environment cues on perceived merchandise value and patronage intentions. J. Mark. **66**(2), 120–141 (2002)
6. Ballantyne, D., Frow, P., Varey, R.J., Payne, A.: Value propositions as communication practice: taking a wider view. Ind. Mark. Manage. **40**(2), 202–210 (2011)
7. Bellizzi, J.A., Hite, R.E.: Convenience consumption and role overload convenience. J. Acad. Mark. Sci. **14**(4), 1–9 (1986)
8. Bentes, A.V., Carneiro, J., da Silva, J.F., Kimura, H.: Multidimensional assessment of organizational performance: integrating BSC and AHP. J. Bus. Res. **65**(12), 1790–1799 (2012)
9. Berry, L.L., Seiders, K., Grewal, D.: Understanding service convenience. J. Mark. **66**(3), 1–17 (2002)
10. Crosby, P.B.: Quality is Free: The Art of Making Quality Certain, vol. 94. McGraw-Hill, New York (1979)
11. Day, G.S.: Managing market relationships. J. Acad. Mark. Sci. **28**(1), 24–30 (2000)
12. Drucker, P.F.: The Practice of Management: A Study of the Most Important Function in America Society. Harper & Brothers (1954)
13. Eid, M.I.: Determinants of e-commerce customer satisfaction, trust, and loyalty in Saudi Arabia. J. Electron. Commer. Res. **12**(1), 78 (2011)
14. Eroglu, E., et al.: Factors affecting consumer preferences for retail industry and retailer selection using analytic hierarchy process. Kafkas Universitesi _ Iktisadive _Idari Bilimler Fakultesi Dergisi **4**(6), 43–57 (2013)
15. Fairchild, A.M., Peterson, R.R.: Business-to-business value drivers and e-business infrastructures in financial services: collaborative commerce across global markets and networks. In: Proceedings of the 36th Annual Hawaii International Conference on System Sciences, pp. 10. IEEE (2003)
16. Frankenberger, K., Weiblen, T., Gassmann, O.: Network configuration, customer centricity, and performance of open business models: a solution provider perspective. Ind. Mark. Manage. **42**(5), 671–682 (2013)
17. Galbraith, J.R.: Designing the customer-centric organization: a guide to strategy, structure, and process. John Wiley & Sons (2011)
18. Garvin, D.: Quality on the line. Harv. Bus. Rev. 65–75 (1983)
19. Gebauer, H., Gustafsson, A., Witell, L.: Competitive advantage through service differentiation by manufacturing companies. J. Bus. Res. **64**(12), 1270–1280 (2011)
20. Hanel, T., Felden, C.: The role of operational business intelligence in customer centric service provision (2014)
21. Hart, C.W.: Customers are your business. Market. Manag. **8**(4), 6 (1999)
22. Hemel, C.v.d., Rademakers, M.F.: Building customer-centric organizations: Shaping factors and barriers. J. Creating Value **2**(2), 211–230 (2016)
23. Jones, P., Clarke-Hill, C., Shears, P., Comfort, D., Hillier, D.: Radio frequency identification in the UK: opportunities and challenges. Int. J. Retail Distrib. Manag. **32**(3), 164–171 (2004)
24. Karras, D.J.: Statistical methodology: Ii. reliability and variability assessment in study design, part a. Acad. Emerg. Med. Official J. Soc. Acad. Emerg. Med. **4**(1), 64–71 (1997)
25. Kowalkowski, C.: The service function as a holistic management concept. J. Bus. Ind. Market. **26**(7), 484–492 (2011)
26. Kruskal, W.H., Wallis, W.A.: Use of ranks in one-criterion variance analysis. J. Am. Stat. Assoc. **47**(260), 583–621 (1952)

27. Kumar, V., Reinartz, W.: Creating enduring customer value. J. Mark. **80**(6), 36–68 (2016)
28. Lee, W.B., Lau, H., Liu, Z.-Z., Tam, S.: A fuzzy analytic hierarchy process approach in modular product design. Expert. Syst. **18**(1), 32–42 (2001)
29. Levitt, T.: Marketing Myopia. Boston (1960)
30. Loonam, M., O'loughlin, D.: An observation analysis of eservice quality in online banking. J. Fin. Serv. Market. **13**(2), 164–178 (2008)
31. Lu, P., Lukoma, I.: Customer satisfaction towards retailers ICA, ICA NARA and coop forum. diva-portal. Gotland University (2011)
32. Lumpkin, J.R., Burnett, J.J.: Identifying determinants of stock type choice of the mature consumer. J. Appl. Bus. Res. (JABR) **8**(1), 89–102 (1992)
33. Marcus, C., Collins, K.: Top-10 marketing processes for the 21st century. Gartner Group Report SP-20-0671 (2003). http://www.gartner.com
34. Munasinghe, D., Hemakumara, G., Mahanama, P.: Gis application for _nding the best residetial lands in ratnapura municipil council area of sri lanka. Int. Res. J. Earth Sci. **5**(10), 11–22 (2017)
35. Naik, C.K., Gantasala, S.B., Prabhakar, G.V.: Service quality (servqual) and its effect on customer satisfaction in retailing. Eur. J. Soc. Sci. **16**(2), 231–243 (2010)
36. Ondrus, J., Pigneur, Y.: Coupling mobile payments and crm in the retail industry. In: Proceedings of the IADIS International E-Commerce, Lisbon, Portugal (2004)
37. Osborne, P., Ballantyne, D.: The paradigmatic pitfalls of customer-centric marketing. Mark. Theory **12**(2), 155–172 (2012)
38. Pan, Y., Zinkhan, G.M.: Determinants of retail patronage: a metanalytical perspective. J. Retail. **82**(3), 229–243 (2006)
39. Parasuraman, A., Zeithaml, V.A., Berry, L.L.: A conceptual model of service quality and its implications for future research. J. Mark. **49**(4), 41–50 (1985)
40. Reinartz, W., Dellaert, B., Krafft, M., Kumar, V., Varadarajan, R.: Retailing innovations in a globalizing retail market environment. J. Retail. **87**, S53–S66. 15 (2011)
41. Rosenbaum, M.S., Cheng, M., Wong, I.A.: Retail knockoffs: consumer acceptance and rejection of inauthentic retailers. J. Bus. Res. **69**(7), 2448–2455 (2016)
42. Rust, R.T., Moorman, C., Bhalla, G.: Rethinking marketing. Harv. Bus. Rev. **88**(1/2), 94–101 (2010)
43. Santy, T.: The analytical hierarchy process: planning, priority setting, resource allocation. Decision Making Series, McGraw Hill, New York, USA (1980)
44. Sciglimpaglia, D., Ely, D.: Internet banking: a customer-centric perspective. In: Proceedings of the 35th Annual Hawaii International Conference on System Sciences, pp. 2420–2429. IEEE (2002)
45. Seiders, K., Berry, L.L., Gresham, L.G.: Attention, retailers! How convenient is your convenience strategy? MIT Sloan Manag. Rev. **41**(3), 79 (2000)
46. Shabaz, M., Kumar, A.: Sa sorting: A novel sorting technique for large-scale data. J. Comput. Netw. Commun. (2019)
47. Shah, D., Rust, R.T., Parasuraman, A., Staelin, R., Day, G.S.: The path to customer centricity. J. Serv. Res. **9**(2), 113–124 (2006)
48. Sorescu, A., Frambach, R.T., Singh, J., Rangaswamy, A., Bridges, C.: Innovations in retail business models. J. Retail. **87**, S3–S16 (2011)
49. Srivastava, V.: Passenger car owners' perceptions of after sales service quality of service centers: An assessment of the servqual dimensions. J. Bus. Ind. Mark. **4**(1), 97–101 (2012)
50. Srivastava, V.: Assessment of advertisement acceptance of idea, airtel and vodafone telecom services among various age groups in Ghaziabad. Raffles Bus. Rev. **1**(1), 103–107 (2016)
51. Srivastava, V., Tyagi, A.: The study of impact of after sales services of passenger cars on customer retention. Int. J. Curr. Res. Rev. **5**(1), 127 (2013)
52. Tanaka-Ishii, K., Tezuka, S., Terada, H.: Sorting texts by readability. Comput. Linguist. **36**(2), 203–227 (2010)
53. Tauber, E.M.: Marketing notes and communications: why do people shop? J. Mark. **36**(4), 46–49 (1972)

54. Tuli, K.R., Kohli, A.K., Bharadwaj, S.G.: Rethinking customer solutions: from product bundles to relational processes. J. Mark. **71**(3), 1–17 (2007)
55. Vlasic, G., Tutek, E.: Drivers of customer centricity: role of environmental-level, organization-level and department-level variables. Zagreb. Int. Rev. Econ. Bus. **20**(2), 1–10 (2017)

Printed in the United States
by Baker & Taylor Publisher Services